Horse-drawn wide boat, H. Sabey
& Company, Paddington

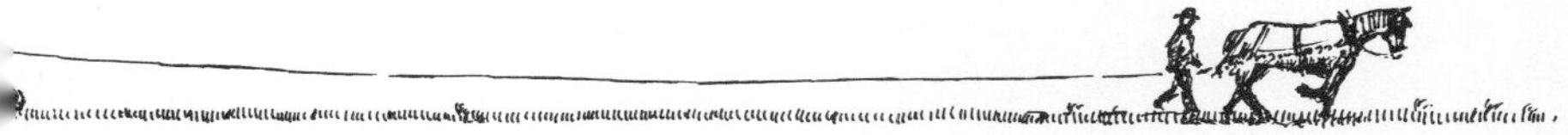

Steam tug *Buffalo*, built 1890

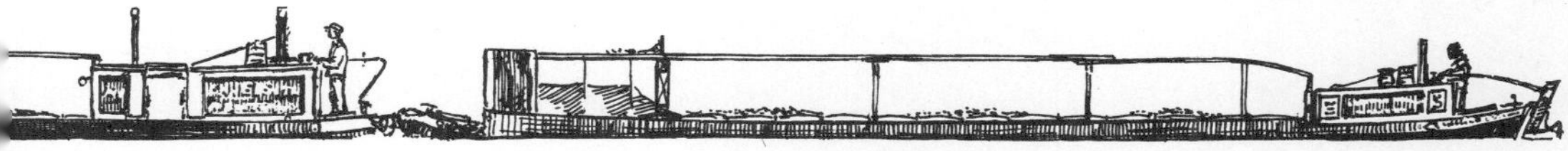

Motor narrow boat *Hasty* and
cross-strapped butty, Emanuel
Smith, Brentford, 1912

THE GRAND JUNCTION CANAL

To Philippa

With best wishes for the future from all at Grand Union Canal (South)

Dec '98

Overleaf: Boats alongside New Mill at Tring, regularly supplied by canal in the 1930s. Latterly the Aylesbury-based Harvey-Taylor fleet was the main carrier, bringing wheat up from Brentford. *Heygates Limited*

THE GRAND JUNCTION CANAL

by

ALAN H. FAULKNER

W. H. WALKER & BROTHERS LIMITED
RICKMANSWORTH 1993

Published by
W.H. Walker & Brothers Limited
ISBN 0 9517923 1 8

British Library Cataloguing-in-Publication Data

A catalogue record for this book is
available from the British Library

Designed by Robert Malster
Printed in Great Britain by
The Wolsey Press, Ipswich, Suffolk.

CONTENTS

To my wife
ANITA

Foreword

Some few months ago I was both surprised and delighted to receive through the post a handsome and beautifully produced book, *Walkers' of Ricky*. This was the first of Anthony Walker's productions, his own history of his family firm of Rickmansworth boatbuilders. It is a book that I treasure because of its quality, the standard of its research and the interest of its illustrations.

Among those illustrations is a lovely profile of a wooden butty boat which was my home and my place of work. She had first slid sideways down into the dark waters of the Grand Union Canal (always dark below the paper mills) from Walkers' yard, below Rickmansworth lock, in 1935.

Warwick, as she became, lived on to experience a kind of apotheosis, with her motor and with a pair of Walker "Ovaltines". All four attended the first national boat rally at Market Harborough and became the cynosure of all eyes with their glittering paint and brass, heavy lace, scrubbed ash gunwales and decorative cotton line.

I am honoured to be asked to write a foreword to the second of Anthony Walker's most admirable publications. What better way to celebrate the bicentenary of the Grand Junction Canal than by combining research, illustration and book production in creating such a definitive record ?

It must be hard for anyone nowadays to understand the overwhelming impact that the life of the canals had on someone joining them in almost total ignorance of what this life entailed. To the women crews of the Second World War, and we do get a mention in Alan Faulkner's account, it was somewhat like travel in a space machine.

Black Jack, Coppermill, Springwell, Stockers, Ricky, Walkers', Croxley, Casey (Cassiobury), the wonderful litany of lock names carrying one up to Cowroast and over the summit. The smells, the sickly sweet foaming water of paper-mill washings; pungent lime-juice barrels; the heady gaseous coking smells of Longford and the bottom road. The dark flight of starlings over the Tring reservoirs and the constant slow disclosure through bridges and in reflections of the ever-changing countryside. None of these things can now be quite the same; as far as I am concerned they were the clothing of history.

A beautiful job has been done by everyone associated with the making of this book, and I commend it to a generation of readers who can never experience the sights and smells, that bygone way of life, at first hand.

Stanley Pontlarge — Sonia Rolt

Preface

The Grand Junction was one of the country's leading canals; as it linked the capital with the industrial heartland it could hardly be anything other than successful. Its main line was over ninety miles long, with more than 100 locks, and its construction represented a major engineering achievement in its day. The greater part was built in well under seven years, and boats were able to pass from one end to the other in under twelve years of the work having started. For a time the Grand Junction was probably the largest joint stock company in the world.

On 30 April 1993 the canal, now an important constituent of the present-day Grand Union Canal, celebrates the bicentenary of the passing of its Act, and it is to mark this anniversary that this book is being re-issued. It was originally published in 1973; the text has now been completely revised and expanded, and many more illustrations have been included. This coincides with the launch of the British Waterways Board's "Canals 200" campaign to highlight the start of the so-called Canal Mania in 1793.

My own introduction to the canal was in the summer of 1953 when my parents discovered a picnic place beside a canal bridge at Cheddington. To a small boy it was ideal, as on the other side of the bridge was a railway line bearing a constant procession of steam trains. The canal bridge was another source of interest as it was, and still is, of the true humped-backed variety, and vehicles had to negotiate it cautiously, with much hooting of horns. But what became the biggest attraction was the canal itself and the pairs of narrow boats, which were continually passing by.

I returned to this spot many times, and then began exploring further afield; from the great windswept embankment across the Great Ouse valley at Wolverton to the seclusion of the deep cutting at Tring through the Chiltern hills; from the deserted waters of the Aylesbury branch to the bustle of busy Brentford depot; from the mysterious and enclosed world of Blisworth tunnel to the wide-open waters of the Tring reservoirs; from the lumbering barges coming off the Thames at high tide to the cluster of idle narrow boats laid up in a backwater at Braunston.

This interest, once kindled, led to me wanting to find out how the canal came into being, and about the history of the boats. In turn this led on to my first article—about the Newport Pagnell Canal—in 1961 in *The Bucks Standard* newspaper. The story of other branches followed, and eventually a publisher suggested a book to cover the whole canal.

Since those early days the whole character of the canal has changed; from being an important artery of commerce the waterway has been turned into a major leisure amenity. In 1953 pleasure craft were starting to make their appearance, but they were few and far between. Industry, however, was fast turning its back on the canal and using supposedly more modern and efficient forms of transport, and the number of working boats and barges was declining. At one time there were fears for the very survival of the canal, but with the growth of recreational use the canal is as busy today as it ever has been. This book traces the history of the canal from its origins, through its heyday as a commercial carrier, to its present-day role.

Colchester Alan H Faulkner

Acknowledgements

I have reason to be grateful to many people who have assisted me over the years in piecing together the history of the Grand Junction Canal. During my main research period the company's minute books were housed at the office of the British Transport Historical Records near Paddington, where the archivist and his staff were most helpful during my frequent visits. I am also grateful to the staffs of the Northampton Central Library, the British Museum Newspaper Library, the House of Lords Record Office, and to the county archivists for Bedfordshire, Buckinghamshire, Hertfordshire, Northamptonshire and what was then Greater London, all of whom have helped me in various ways.

I owe a great debt to the late Charles Hadlow, the first curator of the Canal Museum at Stoke Bruerne, who put his entire set of notes, built up over a lifetime of service on the canal, at my disposal and who really started me on the road to this book. His successor at the museum, Richard Hutchings, also went out of his way to be helpful, and more recently I have been greatly assisted by Tony Conder, and particularly by his archive assistant Roy Jamieson, at the National Waterways Museum at Gloucester. Many other officers of the British Waterways Board have also helped me over the years.

Many fellow members and friends from the Railway & Canal Historical Society have contributed their knowledge, skills, help and encouragement over the years. Sadly not all are now with us, but I record my particular gratitude to Hugh Compton, Michael Ewans, Charles Hadfield, Alan Richardson, Philip Stevens, Philip Weaver and Geoffrey Webb. And my thanks to so many others—too many to mention by name—who have contributed, in greater or lesser degree, to this project.

Each of the photographs used in the book is acknowledged individually in the caption, but it will become obvious I have drawn heavily on the large collection at the National Waterways Museum. Most of the the maps are based upon the Ordnance Survey maps, and have been redrawn with the sanction of the controller, Crown copyright reserved.

I am particularly pleased that Mrs Sonia Rolt has kindly contributed a Foreword to this new edition. To many she needs no introduction as she has been actively involved with the canals for many years, being one of the small group of women recruited in the Second World War to help crew the boats.

Finally I must express my sincere thanks to my new publisher, W H Walker & Brothers Ltd, and to its managing director, Anthony Walker. Both have family connections going back over 150 years with the canals, and over 100 years with the Grand Junction in particular, and I am grateful to them for agreeing to republish this book. My sincere thanks also to my support team comprising Edward Paget-Tomlinson, who has been responsible for most of the maps and drawings, and Robert Malster, who has had the difficult task of putting everything together.

The elegant balustraded bridge at the Grove, Cassiobury Park, Watford, was built in this way in deference to the Earl of Clarendon.

The Canal is Born 1

IN THE eighteenth century Birmingham and the West Midlands developed into one of the biggest manufacturing centres in this country. The area depended heavily on the transport facilities provided by the new generation of inland waterways[1], both for the carriage of raw materials and the despatch of finished products. As a large proportion of the trade was with London the need for a canal to link the two centres became more and more imperative.

In July 1790 a somewhat circuitous route had been opened. From Birmingham it involved the Birmingham & Fazeley Canal, most of the Coventry Canal, the Oxford Canal, and the river Thames from Oxford down to London. The distance was over 230 miles from the centre of Birmingham to London Bridge, whereas they are only 105 miles apart in a direct line. There were also over 100 locks, numerous navigation weirs on the Thames, and the river was generally in poor condition.

The first moves to create a shorter route came towards the end of 1791. A canal was suggested from the Oxford at either Braunston, near Daventry in Northamptonshire, or Aynho, further south in Oxfordshire, to the Thames at Isleworth in Middlesex[2].

The need for an improved link was underlined by the large increase in trade once the through route was opened. This came not only from the Midlands but from the Potteries and the North West. Trade was, however, handicapped by the roundabout route through Oxford and the poor condition of the Thames navigation. The distances involved from either Braunston or Aynho would be much shorter, and the canal would bypass most of the Thames.

Other schemes were coupled in with the Braunston scheme, which was known as the Braunston Canal, and later as the Grand Junction Canal. A canal from the river Soar at Leicester through Market Harborough to the river Nene at Northampton was promoted early in 1792. It was intended to link up with the Grand Junction by a branch, thus giving an improved through route between London and the East Midlands. Similarly two canals were planned to improve the route between Birmingham and Braunston. The Stratford-upon-Avon Canal was also promoted at this time.

The Grand Junction scheme was first made public in April 1792, when the *Northampton Mercury* had a long and mostly favourable review[3]. An initial survey, commissioned and paid for by the Marquis of Buckingham[4], had been carried out earlier in the year by James Barnes of Banbury, who had previously worked on the Oxford Canal. The Marquis was acting on the most disinterested of motives, being concerned simply to try and improve communications in the area.

Opposite page: Thomas Clayton tar boats and other craft waiting at the eastern end of Braunston tunnel for the steam tug in August 1907. The tug which will take them through the tunnel is approaching in the background. *Northampton Libraries*

William Praed, born in 1749, served as the Grand Junction's first chairman from 1 June 1793 to 9 February 1821. He died at Trevethoe in Cornwall on 9 October 1833 at the age of 84. *National Waterways Museum*

On 20 July a public meeting was held at the Bull Inn, Stony Stratford in Buckinghamshire. So many people attended that the business had to be conducted in the parish church, the only place large enough to hold them all. Barnes' survey was presented and enthusiastically approved, and the decision was taken to apply for an Act of Parliament[5]. A committee was set up and the subscribers chose William Praed, MP, as their chairman and leader. It was a fortuitous decision, as he was to prove a tower of strength.

Praed was born in 1749 and followed his father into banking. In 1778 he married Elizabeth Tyringham, another banker's daughter, and went to live at Tyringham Hall near Newport Pagnell in Buckinghamshire. He was Member of Parliament for St Ives in Cornwall from 1781 to 1806 when he resigned to concentrate on banking. He founded his own firm, Praed & Co, bankers of Fleet Street, in 1801.

The meeting appointed Edward Oakley Gray, of Buckingham, and Acton Chaplin, of Aylesbury, as solicitors and clerks to the project, with Philip Box of Buckingham as treasurer. A small steering committee was set up to assist the main committee. A subscription list was opened, the initial capital being set at £350,000 in £100 shares. No one was to hold more than ten shares. An immediate call of £1 per share was made to help pay for the expenses of further surveys and the costs of obtaining an Act.

Landowners whose property would be crossed by the canal were encouraged to support the scheme by giving them an option to subscribe for one share for every furlong of their land used, up to the maximum of ten.

Before applying for an Act it was essential for Barnes to draw up a detailed plan and section of the proposed line. William Jessop, one of the leading civil engineers of the day, was called in to examine and verify Barnes' plans.

Statutory notices for the Grand Junction were published at the beginning of September in the principal newspapers covering its line. These showed the main canal running from Braunston to Long Buckby, Weedon, Blisworth, Cosgrove, Linford, Fenny Stratford, Linslade, Marsworth, Tring, Berkhamsted, Hemel Hempstead, Watford, Ruislip, Harrow, Southall and Hanwell to Brentford. Branch canals to Daventry, Stony Stratford, Uxbridge and Northampton were included.

A second notice was issued which showed a variation of the canal south of Watford passing through Rickmansworth, Denham, Uxbridge and Harlington to Southall. It included all the branches, except that to Uxbridge.

The promoters were obviously undecided on the route south of Watford. Barnes' original plan was for a direct line through Harrow, whereas the alternative was somewhat roundabout as it followed the Colne valley through Uxbridge.

The decision was influenced by a resolution passed at a meeting held in Watford on 20 October[6]. This was organised in support of

Opposite page: The routes of the rival Grand Junction and Hampton Gay schemes, with other canals either existing or proposed in 1792.

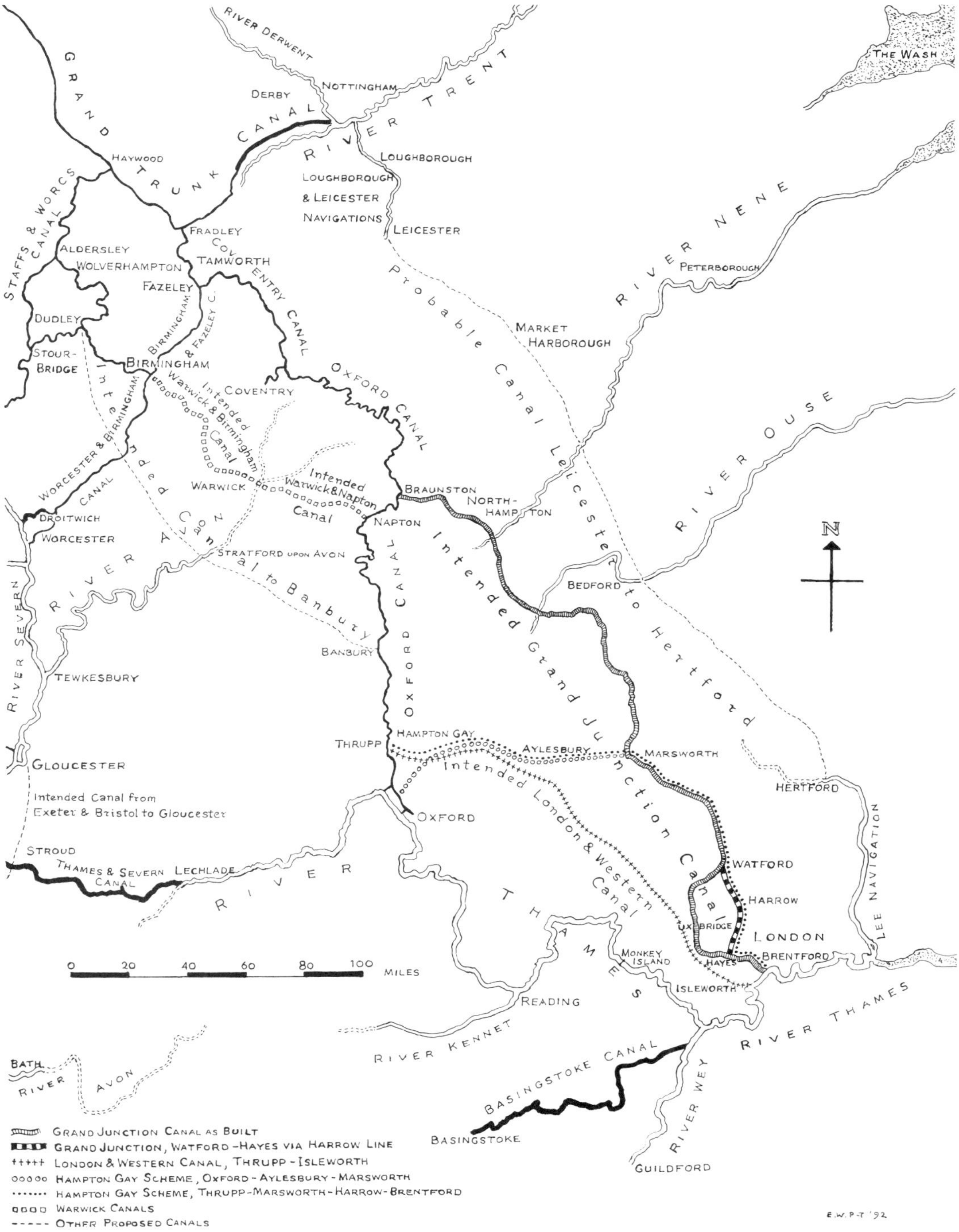
River Derwent
Nottingham
Derby
River Trent
Grand Trunk Canal
The Wash
Haywood
Loughborough
Loughborough & Leicester Navigations
Leicester
Staffs & Worcs Canal
Aldersley
Wolverhampton
Fradley
Coventry Canal
Tamworth
Fazeley
Dudley
Birmingham & Fazeley C.
Stour-Bridge
Birmingham
Probable Canal Leicester to Hertford
River Nene
Peterborough
Market Harborough
Oxford Canal
Coventry
Intended Warwick & Birmingham Canal
Intended Canal
Worcester & Birmingham Canal
Intended Warwick & Napton Canal
Warwick
Braunston
North-Hampton
River Ouse
Droitwich
Worcester
Napton
Intended Canal to Banbury
Stratford upon Avon
River Avon
Bedford
N
River Severn
Banbury
Intended Grand Junction Canal
Tewkesbury
Hampton Gay
Thrupp
Aylesbury
Marsworth
Gloucester
Intended London & Western Canal
Hertford
Intended Canal from Exeter & Bristol to Gloucester
Oxford
Lee Navigation
Stroud
Thames & Severn Canal
Lechlade
River Thames
Watford
Harrow
Uxbridge
London
Monkey Island
Hayes
Brentford
Isleworth
0 20 40 60 80 100 Miles
Reading
River Kennet
Basingstoke Canal
River Thames
Bath
River Avon
River Wey
Basingstoke
Guildford
Grand Junction Canal as Built
Grand Junction, Watford-Hayes via Harrow Line
London & Western Canal, Thrupp-Isleworth
Hampton Gay Scheme, Oxford-Aylesbury-Marsworth
Hampton Gay Scheme, Thrupp-Marsworth-Harrow-Brentford
Warwick Canals
Other Proposed Canals
E.W.P-T '92

the project by the Earls of Clarendon and Essex. Both Jessop and Barnes attended to explain the merits of the two routes. The meeting opted for the Uxbridge route and also for a branch to be built to Watford, as the line was to pass this important town some distance to the west.

Five days later at a general meeting of the subscribers in Northampton the Uxbridge line was chosen[7]. Jessop reported his approval of Barnes' proposed route, merely recommending cutting off a bend near Leighton Buzzard and moving the southern terminus slightly downstream to where the river Brent joined the Thames. He stressed that the main works would be at the three tunnels at Braunston, Blisworth and Langleybury, and at the Tring cutting. He also warned of possible problems in supplying the two summit levels with water, and particularly that at Tring.

Jessop estimated the cost of the main canal at £372,175. It would have a top width of 42ft, a bottom width of 28ft and a depth of 4½ft. The locks were to be 14½ft wide and 80ft long to take Thames or Trent barges. The Daventry branch would cost a further £6,000 and that to Northampton £18,785. By this time the subscription list totalled £361,900.

The Oxford Canal Company was totally opposed to the plan as it would lead to much of the through traffic being lost to its canal. Its answer was to promote a canal to bypass the Thames, but using as much of its line as possible for the London trade. This was the London & Western Canal, later known as the Hampton Gay Canal[8], which was surveyed by Samuel Simcock and Samuel Weston. The sixty-mile line was to run from Hampton Gay, some six miles north of Oxford, to Marylebone. Other canals were planned to strengthen the Oxford's position, one of them to run from the Stratford-upon-Avon Canal at Stratford to Cropredy on the Oxford, just north of Banbury.

One of the first public references to the Hampton Gay scheme was on 24 March 1792 when the *Northampton Mercury* mentioned the plan briefly. Later, in an editorial[9] weighted heavily in support of the rival Grand Junction scheme, it stressed the difficulties and hoped the promoters would have second thoughts. The paper claimed that many of the Oxford's shareholders favoured the Grand Junction project anyway.

At a meeting in Banbury on 29 August the Hampton Gay promoters decided to apply to Parliament for an Act[10]. The statutory notices, which appeared just after those for the Grand Junction, showed a choice of three routes: from Shipton by way of Aylesbury, Wendover, Amersham, Uxbridge and Harrow to Marylebone, with a branch from Uxbridge to the Thames at Isleworth; from Godstow near Oxford to Aylesbury and then to Marsworth to link up with the Grand Junction; and from Shipton to Aylesbury and Marsworth and thereafter on a similar course to the proposed Grand Junction line through Watford and Harrow to the Thames at Isleworth.

It was unlikely there would be sufficient trade for both the

Grand Junction and the Hampton Gay and the two committees had met to try and reach an agreement to link up the two projects. The Marquis of Buckingham had put forward the idea of the junction between the two lines at Marsworth, but it had been rejected by the Grand Junction. Early in November the Grand Junction made a fresh approach, but no compromise could be reached.

At a meeting in Woodstock on 21 November the Oxford proprietors decided by a big majority to back the Hampton Gay by every means in their power, and likewise to oppose the Grand Junction[11]. On 17 January 1793 both the Oxford and the Hampton Gay proprietors petitioned against the Grand Junction scheme[12].

Despite this the Grand Junction had powerful backing, with many prominent landowners, such as the Duke of Grafton and Earl Spencer, not only supporting but in some cases actively promoting it. So much so that it triumphed over the Hampton Gay project, which was then withdrawn. This still left the Oxford in opposition, and the Grand Junction eventually had to agree to an annual £10,000 income guarantee.

William Jessop was chief engineer to the Grand Junction whilst it was being built. Born at Devonport in 1745, he died at Butterley Hall on 18 November 1814, aged 68. *National Waterways Museum*

Negotiations also took place to lease the northern part of the Oxford between Braunston and Hawkesbury. The Grand Junction was planned as a wide, or barge, canal. The promoters wanted to widen the northern part of the Oxford to barge standards to improve access to Coventry and the Warwickshire coalfields and, it was hoped, further north in time. The two sides failed to agree, but this was the start of a long campaign by the Grand Junction for other companies to widen their waterways and establish a barge route from London to the Midlands and the North West.

Once the Oxford had been bought off there was little opposition to the Act, which had its third reading in the House of Commons at the end of March[13], was passed by the House of Lords committee at the end of April[14], and received the Royal Assent on 30 April 1793[15]. The ambitious project was under way.

Under the Act 417 shareholders were incorporated as "The Company of Proprietors of the Grand Junction Canal". Among them were a duke, a marquis and five earls. They had power to raise £500,000 in £100 shares, with a further £100,000 should it be needed. To finance the business so far a second call of £2 per share had been made in December 1792, a third of £2 at the end of April 1793, and another of £5 in June[16].

Various tonnage rates were authorised. Lime and limestone could be charged 0.1p (one farthing) per ton mile; livestock, bricks, slate, tiles, sand, manure, pig iron, pig lead and iron stone 0.2p (one halfpenny) per ton mile; coal and coke 0.3p (three farthings) per ton mile; and all other goods 0.4p (one penny) per ton mile. There was an additional toll to the City of London of 0.2p per ton on all goods passing between the canal and the Thames. Fractions of a mile were charged as one mile; fractions of a ton were charged to the nearest quarter below.

Various exemptions were granted. The owners of wharves on

the lower section of the river Brent, which was to be used for the canal, paid no tolls at all, even to the City of London. Troops, their horses, arms, equipment and stores were all to be carried toll free. Sand and gravel for making and repairing public roads were toll free provided they did not pass through any locks. The same applied to manure for the land.

The Act provided that if the tolls to the City of London, which controlled the Thames up to Staines at the time, were less than £200 up to 24 June 1795 the company had to make good the deficiency. The amount increased by £100 each year until 1801 when the annual sum was £1,000. This was the price enforced by the City for the new canal joining the river.

The agreement with the Oxford was also written into the Act, special tolls being authorised on all goods passing between the two canals. All coals from the Oxford to the Grand Junction were charged 14p per ton and all other goods 21½p. Once the Grand Junction was opened from Braunston to Old Stratford any shortfall in these tolls below £5,000 each year had to be made good. Once opened throughout, or after 1 January 1804, the guarantee rose to £10,000.

Another section of the Act dealt with the appointment of commissioners to settle any differences and assess damages and compensation caused by the canal works. The commissioners could have no direct or indirect interest in the canal. They were drawn from landowners owning property at least £100 in value in the five

A narrow boat belonging to the cement maker Charles Nelson & Co Ltd of Stockton entering Denham lock, the deepest lock on the canal. It was undergoing repair at the time. *Railway & Canal Historical Society*

counties through which the canal passed, and certain other responsible people in those counties. If their efforts were unsuccessful the Act provided for the case to be tried by a jury.

Many other restrictions and stipulations were laid down in the Act. Wharfage dues were set out; precise regulations for taking water from rivers and streams were specified; in places the towing path had to be on certain sides of the canal so as not to interfere with specified property; the opening hours at various times of the year were set out; and there were rules for those using the canal and particularly covering the use of the locks.

The plans deposited with Parliament in support of the Act showed the main line to be just over ninety miles long, with 121 locks. In addition there were to be branches to Daventry, Northampton, Stratford and Watford.

The south-east portal of Braunston tunnel.

From the Oxford Canal at Braunston a flight of locks was to carry the canal up to the summit level, where there was a tunnel through a ridge of hills. The descent from the summit commenced at Norton through another flight of locks. There was then a long level stretch to Stoke Bruerne, through a second tunnel at Blisworth. The canal then dropped down into the valley of the river Great Ouse, which was crossed near Wolverton.

After the climb out of the valley there was another long level stretch to just south of Fenny Stratford, where the climb to the highest point on the canal at Tring in the Chiltern hills started. Southwards from the Tring summit the canal ran parallel with the rivers Bulbourne, Gade and Colne through Berkhamsted, Hemel Hempstead, Watford and Uxbridge, except for a detour at Langleybury where there was a third tunnel. From Uxbridge there was another level stretch to Norwood and Hanwell, where a flight of locks dropped the canal down to the river Brent. Its course was then followed through Brentford to the Thames.

NOTES TO CHAPTER ONE

1 Hadfield, Charles. *Canals of the East Midlands*, 1966.
2 Oxford Canal Committee minutes, 11 January 1792.
3 *Northampton Mercury*, 14 April 1792.
4 Ibid, 21 July 1792.
5 Ibid, 27 October 1792.
6 Ibid, 3 November 1792.
7 Ibid, 27 October 1792.
8 Compton, Hugh. *The Oxford Canal*, 1976.
9 *Northampton Mercury*, 14 April 1792.
10 Ibid, 8 September 1792.
11 Ibid, 1 December 1792.
12 Ibid, 19 January 1793.
13 *Derby Mercury*, 4 April 1793.
14 *Northampton Mercury*, 20 April 1793.
15 33 Geo III, c 80—30 April 1793.
16 *Northampton Mercury*, 27 April 1793.

24-26
THREE LOCKS
SOULBURY
BRAUNSTON
GRAND JUNCTION CANAL
Chiltern Hills
LEIGHTON LOCK
LINSLADE
27
LEIGHTON
Wharf
BUZZARD
GROVE LOCK
28
CHURCH LOCK
29
SLAPTON LOCK
30
31
HORTON LOCK & Wharf
32
IVINGHOE LOCKS
33
SEABROOK LOCKS
34
35
MARSWORTH TWO LOCKS
36
AYLESBURY BRANCH
16 15 14
9-13
1-8
37-38
AYLESBURY
MARSWORTH FLIGHT
39-45
BULBOURNE Workshops
WENDOVER ARM
Wharf
Cutting
TRING
HALTON
WENDOVER
COWROAST LOCK
46
47/8
DUDSWELL TWO LOCKS
49
NORTHCHURCH TOP LOCK
50
NORTHCHURCH LOCKS
51/2
BERKHAMSTED LOCKS
53
54/5
SEWERAGE LOCK, BOURNE END
58
59
BOURNE END BOTTOM LOCK
62 BOXMOOR TOP LOCK
TOPSIDE
BOTTOMSIDE LOCKS
BERKHAMSTED 56
BOURNE END 57
56/7
63 FISHERY LOCK, BOXMOOR
WINKWELL LOCKS
60/1
64 BOXMOOR BOTTOM LOCK
APSLEY LOCKS 65/7
65/7 TWO WATERS
NASH MILLS LOCKS
68/9
69A
KINGS LANGLEY LOCK
HOME PARK MILL LOCK
70
HOME PARK FARM LOCK
71
HUNTON BRIDGE LOCKS
72/3
LADY CAPEL'S LOCK
74
CASSIOBURY PARK LOCKS
75/6
IRONBRIDGE LOCK
77
CASSIOBRIDGE LOCK
78
WATFORD
N
CROXLEY
RICKMANSWORTH
79 COMMON MOOR LOCK
RICKMANSWORTH BRANCH
80 LOT MEAD LOCK
81 BATCHWORTH LOCK
SPRINGWELL LOCK 83
82 STOCKER'S LOCK
84 COPPER MILL LOCK
TROY CUT
HAREFIELD
BLACK JACK'S LOCK
85
WIDEWATER LOCK
86
DENHAM
87 DENHAM DEEP LOCK
88 UXBRIDGE LOCK
UXBRIDGE
Wharves
COWLEY LOCK
89
PADDINGTON BRANCH
Wharves
REGENT'S CANAL
MAIDA HILL TUNNEL
272 YDS
PADDINGTON BASIN
Wharves & Docks
GREENFORD
COWLEY PEACHEY JUNC
SLOUGH
SLOUGH BRANCH
SOUTHALL
Wharves & Docks
BOTWELL
NORWOOD LOCKS
90 91
HANWELL FLIGHT
92-97
Wharves
DAWLEY
BULLS BRIDGE
Docks
OSTERLEY LOCK
98
99 CLITHEROE'S LOCK
GROSVENOR CANAL
KENSINGTON CANAL
100
101
BRENTFORD GAUGING LOCKS (DOUBLE)
THAMES LOCK (TIDAL)
BRENTFORD, Dock
Wharves
ISLEWORTH
HALF TIDE LOCK
RICHMOND
THAMES ESTUARY
BOURNE END
MARLOW
COOKHAM
OXFORD
BOULTER'S LOCK
MAIDENHEAD
RIVER
MONKEY ISLAND
WINDSOR
DATCHET
THAMES
STAINES
TEDDINGTON
WEYBRIDGE
5 0 5 10
MILES
E.W.P-T '92

Building the Main Line 2

Once the Act was obtained the proprietors held their first general meeting on 1 June 1793 at the Crown & Anchor Tavern in London's Strand. William Praed was elected chairman and a presentation of plate was made to him for his work for the canal[1]. A vote of thanks was given to the Marquis of Buckingham who had given such strong support to the project. He was described as the "Projector and Patron of the undertaking", and his coat of arms was incorporated into the company's official seal.

The Grand Junction's committee seal, based on the arms of the Marquis of Buckingham, one of the canal's leading supporters. *National Waterways Museum*

The meeting decided to issue £450,000 worth of shares and an immediate 10% call was made. Edward Gray and Acton Chaplin were confirmed as clerks and Philip Box as treasurer. A committee of thirty-one was appointed including William Praed, the Marquis of Buckingham and such influential people as the Duke of Grafton, the Earl of Essex, the Earl of Clarendon, Earl Spencer and the Hon. Edward Bouverie. Two district committees were formed, one to superintend the canal from Brentford to Marsworth, the other the rest of the line to Braunston.

The main committee met two days later. James Barnes was appointed resident engineer at £2.10 per day and 53p per day expenses. He was to devote his entire time to the canal. Two days later William Jessop was appointed chief engineer and instructions were given to set out the whole line of the canal.

On the administration side Thomas Homer, who had been joint clerk to the Coventry Canal, was appointed on 29 September to "Control, Audit and Methodise the Accounts of the Company" and to superintend matters generally. His salary was £500, with a further £100 for a house until this could be provided for him by the company.

Work started at Braunston and Brentford immediately the Act was obtained. By the middle of May there were 370 men already on the payroll and more were arriving every day[2]. Cutting had also begun on Uxbridge Moor on 1 May[3] and a start was made at Blisworth tunnel and Tring cutting soon after. At the end of the year 3,000 men were engaged.

Most of the excavation was done by hand. In June an experiment was made with a new machine designed to save most of the manual labour in cutting and removing the earth, but it was not a success. It was the combined efforts of men and horses that dug the Grand Junction.

Meanwhile the plans for a barge route to the North West were still being pursued. In July a meeting was held at Meriden in Warwickshire with those companies likely to be affected, and it was agreed that a wide canal would be beneficial if it could be provided

Opposite page: The southern section of the Grand Junction Canal from Brentford to Leighton Buzzard.

at a reasonable price. Nothing practical came out of the conference but the company persevered in its efforts for some time.

Progress in the South

Starting from the Thames, the work made good progress and the canal was formally opened from Brentford to Uxbridge on 3 November 1794[4].

> That part of the Grand Junction Canal from the river Thames near Brentford, to the town of Uxbridge was opened on the 3d instant for coals and all sorts of merchandise to be navigated thereon; comprizing upwards of twelve miles of this great undertaking. The opening of this part of the Canal was celebrated by a variety of mercantile persons of Brentford, Uxbridge and Rickmansworth and their vicinities, forming a large party, attended by a band of music, with flags and streamers, and several pieces of cannon, in a pleasure boat belonging to the Corporation of the City of London, preceding several barges laden with Timber, Coals and other Merchandize to Uxbridge.

Some work remained to be completed and it was not until May 1795 that toll collectors were appointed at Brentford and Uxbridge. There was a problem at Cowley where it was planned to cross the river Colne on the level. Disputes with the millers over water rights

An early picture of the Brentford-based barge *Perseverance*, owned by W. Whitehouse & Son, being loaded with bundles of timber laths.
Kodak Museum

led to an aqueduct having to be built to carry the canal over the river at Cowley lock, and this was not finished until autumn 1795, temporary arrangements being made to pass the traffic in the meantime. There were similar difficulties north of Uxbridge, where another aqueduct had to be built over the Colne at Denham lock.

Further problems arose with the road bridge at Brentford, a three-arched brick structure crossing the tidal part of the canal immediately below what was then the last lock. The canal used the course of the river Brent here, and at high tide it was extremely difficult for barges from the Thames to pass through.

Jessop had inspected it in May 1794 and found the middle arch needed raising by 3ft. He suggested a new brick structure with a 50ft span and a 28ft roadway, but on a slightly different alignment. The work would cost £2,830, but this excluded the cost of purchasing the buildings on the land affected and the provision of a temporary structure.

The company was unwilling to spend such a large sum unless the County of Middlesex and the turnpike road trustees would make a substantial contribution. The county declined, as it considered the old bridge was in good order, and the trustees declined as the county had done so. Jessop had to alter the old bridge as best he could.

Between Brentford and Uxbridge there was a major deviation

A horsedrawn narrow boat heads south past the Cassiobridge wharves at Watford. In 1925 a high-level railway bridge was built over the canal between this spot and the lock in the background. *Hugh McKnight*

Excavating tools—barrow, pick, shovel &c.

from the deposited plans. From Bulls Bridge at Southall, at what is now the junction with the Paddington branch, the canal was to have run due west, passing just north of Cranford Park and Harlington to rejoin the present line just north of West Drayton. Difficulties in obtaining the land resulted in the adoption of a more direct line further to the north, passing through Botwell and Dawley. This deviation involved some heavy earthworks, with a long cutting at Dawley, but it saved over three-quarters of a mile of canal.

Once the canal was open to Uxbridge the company purchased land for a wharf in the commercial part of the town, and in April 1796 a short branch canal and a turning basin were built to serve it.

North of Rickmansworth there was another major deviation from the plans. The line was intended to pass much farther to the west, with a flight of sixteen locks taking the canal up to Langleybury, where a 900yd tunnel was to be driven through the high ground. From the north end of the tunnel a level four-and-a-half-mile stretch was planned, crossing the river Gade on an aqueduct at Kings Langley and rejoining the present line at Two Waters.

In December 1793 a contract was awarded to Charles Jones and John Biggs to construct the tunnel at Langleybury. The work was to be completed by 1 June 1796. Little was done, however, for shortly afterwards a revised line, avoiding the tunnel, was proposed, and in March 1794 negotiations were opened with the landowners who would be involved. In May Barnes was told to stop such work as had begun on the tunnel and build a railway over the high ground at Langleybury instead. A month later a plan was produced to divert the line eastwards to follow the river Gade and avoid the high ground altogether[5]. This involved passing through Cassiobury and Grove parks.

The company calculated it could offer £12,000 to the Earls of Essex and Clarendon, who owned the parks, for their permission, and still show a saving on the costs of the tunnel. William Praed was asked to consult their lordships. He reported back in July that Lord Essex had turned down the offer. Jessop was then asked to do a more detailed estimate of the savings that could be made.

In the autumn of 1794 the company applied to Parliament for leave to bring in an Act to authorise the diversion in the hope an agreement could be made with Lord Essex, who was still on the committee at the time. In November he settled for £15,000 for compensation and for the land needed. Then in March 1795 the Earl of Clarendon agreed to accept £5,000 as compensation.

The Act[6] authorising the diversion was obtained in March 1795, but it was not until November that Barnes set out the line through the two parks. In the following month the company acquired the Cassiobury land and work started early in 1796, but it seems the land at Grove was never purchased by the company.

Amongst the clauses in the Act was one authorising an additional toll of 1p per ton on all goods passing on the diverted line in view

FMC's *Hanwell* passes through Grove ornamental bridge near Watford in 1934. The bridge was built in this style to appease the Earl of Clarendon, who owned the surrounding land. *Birmingham Central Library*

of the safer and speedier passage. Another provided for the canal through the parks to be made as ornamental as possible; hence the decorative bridge at Grove and the fine stretch of water through the park at Cassiobury.

Major delays had been caused north of Uxbridge by the nature of the ground, mainly sand and gravel, through which the canal was driven. Water tended to seep into lock chamber excavations and to drain out of the pounds. Pumps had to be provided at all the locks between Cowley and the Tring summit to keep them clear of water whilst they were being built, and piles had to be driven and sheeting put in to consolidate the loose soil during the work. Similarly many miles of canal in the area had to be given additional lining to make the banks completely watertight.

In November 1796 Barnes was told to concentrate on the section from Uxbridge to the Tring summit to get this stretch opened as soon as possible. By the middle of September 1797 the canal had been completed to Kings Langley[7], twenty-six miles from the Thames. Two months later Barnes reported that all the cutting from Kings Langley to Two Waters, and all the brickwork except at one lock and one bridge, had been completed. This section was opened by the end of the year.

By then cutting was under way on the stretch north of Two Waters and all was ready for work to start on several locks. A large quantity of timber had already been brought in and four million bricks had been stockpiled in strategic places along the route.

A problem arose when land was required across Boxmoor Common. This land was held by forty joint owners and preserved for the benefit of the local people. Whilst there was a clause in the company's Act which permitted trustees to sell land required for the canal, there were doubts if this applied to the joint owners, who otherwise had no powers to dispose of any of the property. The company had what amounted to compulsory purchase powers for land up to 100yd on either side of the line shown in the deposited plans, so presumably the joint owners would feel that they must permit the canal to be built, particularly as it would be of great advantage to the local people.

At a public meeting in 1799 of the people of Hemel Hempstead and Bovingdon a committee of seven from the joint owners was appointed to treat with the company over the price to be paid. Eventually £900 was agreed for the twenty-five acres taken. The money went to build a new wharf on the canal and a new workhouse. There must have been lingering doubts about the legality of the transaction, for when the Boxmoor Trust was formally established by Act of Parliament in 1809 the opportunity was taken to confirm the sale of the land to the company.

In March 1798 Barnes reported that the canal from Two Waters across Boxmoor Common to Berkhamsted was three-quarters complete, but only half finished from Berkhamsted to Tring. Delays were being caused by a shortage of bricks, most of which came from North Hyde, near Southall, and were brought up the canal by

The FMC steamer *Empress* and its butty *Wales* are bound probably for the City Road Basin as they work through one of the Northchurch locks, near Berkhamsted. *Black Country Museum*

barge. Four barges were at work, but they were not enough, and efforts to get more into service had failed. Barnes hoped the position would improve as soon as men and teams of horses were released from farms where they were occupied with the spring sowing.

It had been expected that the canal would be open to Berkhamsted by the beginning of May and to the summit by July. This was now impossible unless a very high and uneconomic rate was paid to attract men and horses to enable the extra barges to be put into service. On the run from North Hyde three teams of horses were needed for each barge, together with the men to tend them. As a result it was not until the autumn of 1798 that the canal was open to Berkhamsted. The Tring summit, thirty-five miles from the Thames, was reached early in 1799.

The engineers faced a formidable problem of how to carry the canal over the Chiltern hills. Eventually it was decided to drive a deep cutting through the hills near Tring, a major undertaking. The cutting is over 30ft at its deepest and extends for over one and a half miles. Disposing of the excavated material presented considerable problems and much was dumped on to the land on either side of the cutting, making it appear deeper than it really is.

Puddling tools—spade and scoop.

Apart from the sheer magnitude of the work the engineers encountered difficulties when both sides of the cutting started slipping. The problem was aggravated by several small springs discovered during the digging. The problem was solved by driving small tunnels, or headings, into the affected areas to drain the water away, and stabilizing the sides of the cutting with timber piles, which cost an extra £11,000.

Work on the three-mile summit and the associated feeder branch to Wendover was started early in the summer of 1793. Both were finished by early 1797. At the time they were isolated from the rest of the line, but as this was a vital section which had to supply water to the canal on both sides of the summit priority had to be given to completing the work as early as possible.

Northwards from the summit there were relatively few difficulties. By November 1797 the first two miles through Marsworth to Cheddington were nearly complete. This length included nine locks. Barnes had started on this section as plenty of good clay had been found which was ideal for brickmaking. A large stock of bricks was being built up, ready for the next summer, when they could be transported by canal southwards to the three-mile stretch to Berkhamsted that was still being built, or northwards as required, achieving a great saving in transport costs.

Another problem arose at Fenny Stratford where the company wanted to build a bridge to carry Watling Street over the canal. The river Ousel was not far away but at a much lower level, which meant that the gradient from the river bridge to the new canal bridge would have been too steep for coaches and other horse-drawn vehicles[8].

Tring cutting, parallel with the railway cutting, with Marshcroft bridge.

In March 1800 a start on the canal bridge had to be deferred until Barnes could reach an agreement with the turnpike trustees. Finally it was decided to divert Watling Street away from the old river bridge to a new three-arched one, and in mid-June Barnes was ordered to complete this bridge as quickly as possible.

In the meantime the canal was opened from the Tring summit to Fenny Stratford on Wednesday, 28 May 1800[9]. The occasion was marked by a number of boats starting from Tring in the morning and arriving in grand procession at Fenny, with the Marquis of Buckingham and other notables attending. There was now a fifty-five-mile stretch open from the Thames involving seventy-seven locks.

Between Hemel Hempstead and Fenny Stratford there were few significant variations from the deposited plans, except for minor differences in the siting of some of the locks. At Marsworth, however, the line was planned to pass further to the east of the village, with a flight of ten locks in under half a mile down from the summit. As built the canal passed virtually through the village, and the locks were spread out over a greater distance by reducing the length of the summit level.

Progress from the North

Braunston tunnel was one of the first parts of the canal to be started. Early in June 1793 Barnes arranged a contract with Charles Jones and his partner John Biggs to drive the tunnel at £9.50 per yard. This price was to include all the materials that would be needed. The tunnel was to be finished by 1 January 1797.

In May 1794 Jessop reported that the walls and arches had been started in two places[10], but he was far from impressed by the contractors. The brickyards which had been set up to produce the large quantity of bricks needed were fairly busy, but the number of rejects was high. Enough bricks had been produced for nearly all the locks and bridges between Braunston and Blisworth.

Shortly afterwards, as a result of this report, Jones was dismissed. To avoid delaying the work the company took direct control of his workforce. Then in September Biggs threw in his contract, saying he could not complete it on the agreed terms. The company took over his work as well, but retained him as a supervisor.

By this time four of the locks at Braunston had been started as well as several bridges. About two miles of canal to the east of the tunnel had been dug, whilst the deep cutting at the tunnel entrance was due to be finished by the end of May. The completed part of the canal was being filled with water so that boats could be used to transport bricks to bridge sites and to the locks at Long Buckby.

Further south the long embankment at Weedon across the river Nene valley was well over half complete and wagonways were being laid down to assist in the work. Embankments at Heyford and Bugbrooke were almost complete except for the clay puddling, and the culvert over the Horsestone Brook at Bugbrooke was nearly finished. Then nearly two miles of canal between Gayton and Blisworth had been excavated and this had involved considerable earthworks.

In February 1795 Jessop discovered that in a couple of the shafts sunk along the course of Braunston tunnel the lines used to ensure the excavation underground was proceeding in the correct direction were wrong. Jessop told Hollingsworth, the company's resident engineer at the tunnel, to have them put right and to inspect the lines regularly in future.

In June Jessop found Hollingsworth had failed to rectify the fault, with the result that part of the tunnel had been driven in slightly the wrong direction. Jessop placed the blame principally on Biggs, whose job it was to ensure the excavation underground followed the straight line set out by Barnes on the ground above. Hollingsworth also came in for severe censure; his pay was halved for a while and he narrowly escaped being dismissed. Because of this mistake Braunston tunnel has a double bend in it to this day.

Further problems arose when quicksands were encountered. Additional shafts had to be sunk to enable the 328yd of quicksand to be overcome, this extra work costing £4,800.

The junction of the Oxford and the Grand Junction canals at Braunston in 1923, showing the Oxford's stop lock and toll house in the background. The Oxford's main line originally passed through the bridge on the left on its way south, but in 1834 this section was by-passed and subsequently truncated into a short branch to serve Nurser's boatyard and the Grand Junction's Braunston warehouse. *William Knibbs*

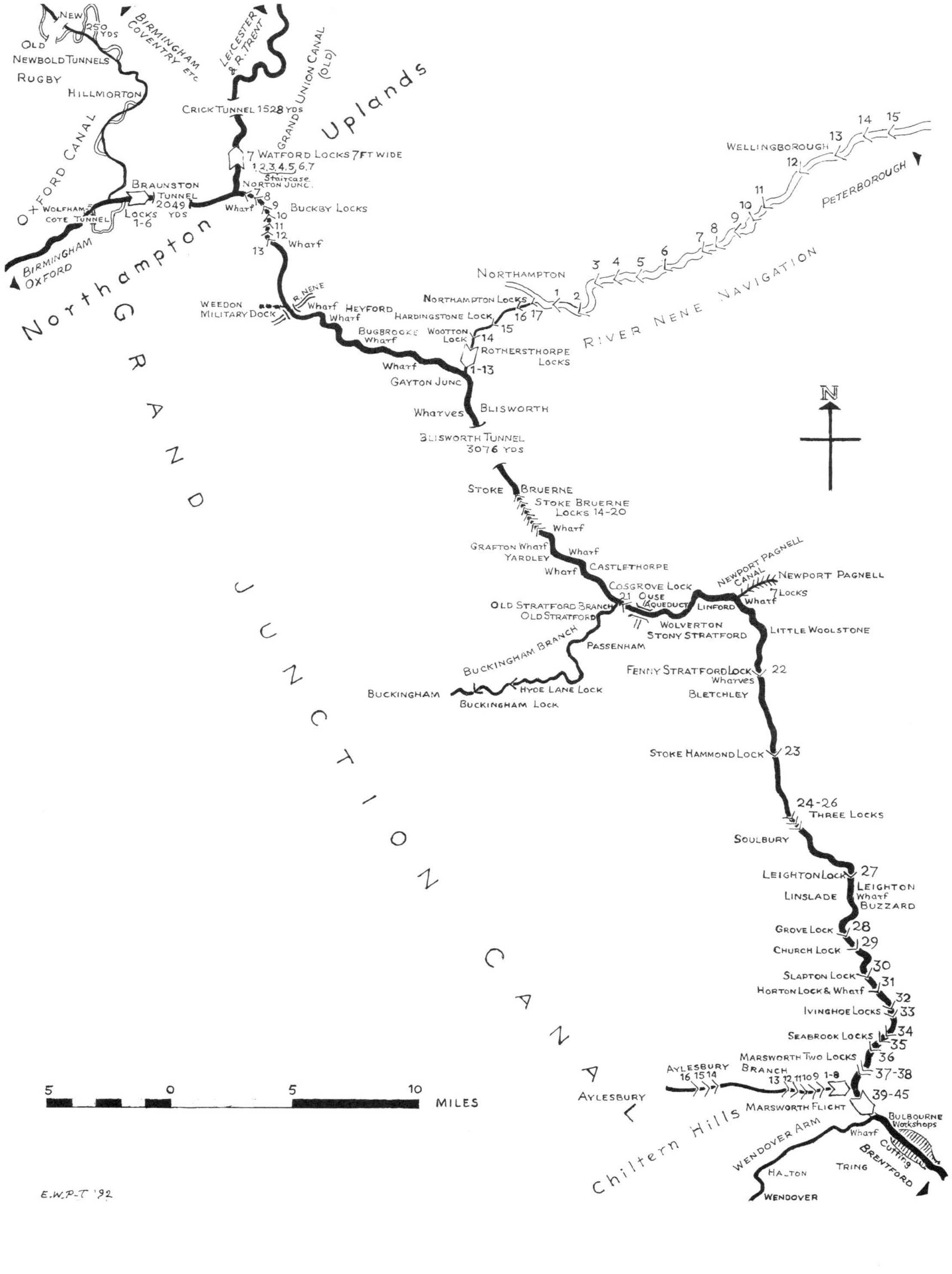
GRAND JUNCTION CANAL
Northampton Uplands
Chiltern Hills
OLD
NEWBOLD TUNNELS
NEW
250 YDS
RUGBY
HILLMORTON
BIRMINGHAM COVENTRY ETC
LEICESTER & R. TRENT
GRAND UNION CANAL (OLD)
CRICK TUNNEL 1528 YDS
7 WATFORD LOCKS 7FT WIDE
1,2,3,4,5 6,7
Staircase
NORTON JUNC.
OXFORD CANAL
BRAUNSTON
TUNNEL 2049 YDS
WOLFHAMCOTE TUNNEL
LOCKS 1-6
BIRMINGHAM OXFORD
Wharf
BUCKBY LOCKS
Wharf
WEEDON MILITARY DOCK
R. NENE
Wharf
HEYFORD Wharf
BUGBROOKE Wharf
Wharf
GAYTON JUNC
NORTHAMPTON
NORTHAMPTON LOCKS
HARDINGSTONE LOCK
WOOTTON LOCK
ROTHERSTHORPE LOCKS
1-13
RIVER NENE NAVIGATION
WELLINGBOROUGH
PETERBOROUGH
Wharves
BLISWORTH
BLISWORTH TUNNEL 3076 YDS
N
STOKE BRUERNE
STOKE BRUERNE LOCKS 14-20
Wharf
GRAFTON Wharf
YARDLEY
Wharf
Wharf
CASTLETHORPE
COSGROVE LOCK
21 OUSE AQUEDUCT
NEWPORT PAGNELL CANAL
NEWPORT PAGNELL
7 LOCKS
Wharf
LINFORD
OLD STRATFORD BRANCH
OLD STRATFORD
WOLVERTON
STONY STRATFORD
LITTLE WOOLSTONE
PASSENHAM
BUCKINGHAM BRANCH
BUCKINGHAM
HYDE LANE LOCK
BUCKINGHAM LOCK
FENNY STRATFORD LOCK 22
Wharves
BLETCHLEY
STOKE HAMMOND LOCK 23
24-26 THREE LOCKS
SOULBURY
LEIGHTON LOCK 27
LEIGHTON Wharf BUZZARD
LINSLADE
GROVE LOCK 28
CHURCH LOCK 29
SLAPTON LOCK 30
HORTON LOCK & Wharf 31
IVINGHOE LOCKS 32 33
SEABROOK LOCKS 34 35
MARSWORTH TWO LOCKS 36
37-38
39-45
AYLESBURY BRANCH
16 15 14
13 12 11 10 9 1-8
AYLESBURY
MARSWORTH FLIGHT
BULBOURNE Workshops
Wharf
Cutting
WENDOVER ARM
HALTON
TRING
BRENTFORD
WENDOVER
5 0 5 10 MILES
E.W.P-T '92

Braunston tunnel was opened on Tuesday, 21 June 1796, and at the same time the canal was opened to Weedon, a distance of almost nine miles from Braunston and involving thirteen locks[11].

One of the original milestones.

> Nothing could be more gratifying than the scene which was exhibited on the opening of the Grand Junction Canal from the Oxford through Braunston Tunnel to Weedon last Tuesday. The wetness of the early part of the morning proved no obstacle to the Committee and Ladies and Gentlemen of the neighbourhood, assembled at Braunston by two o'clock when, preceded by their chairman, Wm. Praed, Esq, M.P., two flags and an excellent band of music, they. . .embarked on board two boats attended by an immense concourse of people. . .the Tunnel, which is 2045 yards long, is the largest and is allowed by professional men to be the best piece of work of any of the kind in the kingdom; and though so many difficulties have attended it, being through quicksands from 30 to 50 feet deep, yet it has been completed in the short space of two years—1500 yards of it having been done in the last nine months.

By now construction of the nine-mile stretch from Weedon southwards to Blisworth with its heavy earthworks was well advanced. It was hoped it would be open early in August[12], but a water shortage prevented the company filling the canal sufficiently for fully laden boats. In the middle of September a new wharf was established at Blisworth[13] by the carriers Roper Barnes & Co for dealing with coal, coke, salt, slates and other commodities, and the canal was opened to it shortly afterwards.

Several deviations were made from the deposited plans in this section. Originally Braunston tunnel was to have been only 1,600yd long, but it was lengthened to over 2,000yd to reduce the deep cuttings at both ends. The Braunston summit was to have been more than four miles long, but was shortened by half a mile due to alterations at Long Buckby, where ten locks were at first proposed in a close flight within five-eighths of a mile. The flight was reduced to seven locks and spread out to start half a mile further north.

On the long level stretch to Blisworth the plans provided for the canal to take a more direct course. As built the contour was followed on several occasions to reduce the earthworks, adding half a mile to the overall length of the canal.

Linking the two sections

There was now a seventeen-and-a-half-mile stretch of canal open from the Oxford Canal at Braunston to Blisworth, but no work had started between Blisworth and Fenny Stratford apart from Blisworth tunnel. Until this section was opened there seemed little point in proceeding southwards, and there was no water to fill the canal anyway. Instead efforts were concentrated on the southern section to Fenny Stratford.

By mid-1800 the cutting from Fenny Stratford to the Great Ouse valley at Wolverton was well under way. In April Jessop reported good progress and forecast completion of the work by the end of the

Opposite page: The northern section of the Grand Junction Canal from Leighton Buzzard to Braunston.

year unless there was a shortage of workmen.

There were several small deviations on this stretch resulting in the canal taking a more roundabout course to avoid extensive earthworks, and there was a major variation on both sides of the Wolverton valley. From the south the last locks to complete the drop down to the Great Ouse were to be between Little Woolstone and Bradwell. After crossing the river on the level the canal was to climb the valley of the river Tove to Stoke Bruerne, the locks being evenly spaced to match the gradual rise beside the river.

Wolverton aqueduct over the Ouse. Originally three brick arches carried the trough, but in 1808 these collapsed and were replaced by cast-iron segments held by two arched members, the new trough being opened in 1811.

Once it had been decided to cross the Great Ouse by means of an aqueduct and embankment (see chapter 3) the canal on both sides of the valley was re-routed to follow the contour and eliminate the proposed locks. The new crossing point was also to be further upstream, where the valley was steeper and narrower and more suitable for the embankment.

Meanwhile it was becoming clear the company had massive problems at Blisworth (see chapter 3) and the tunnel would take a long time to complete. In June 1799 Barnes started work on the canal southwards from the bottom of the proposed flight of locks at Stoke Bruerne to the Wolverton valley. By mid-1800 this work was well advanced with a forecast opening date of 1 September. The line was being filled with water from the river Tove.

Two flights of temporary locks were built across the Ouse valley and a railroad was provided across the top of Blisworth hill to link up the two ends of the waterway. The railroad was completed early in 1801[14], and after a little less than eight years' work the Grand Junction had a through route.

NOTES TO CHAPTER TWO

1 Boyes, Grahame. A Small Token, *Railway & Canal Historical Society Journal,* November 1984.
2 *Northampton Mercury,* 25 May 1793.
3 Lysons, Rev D. *An Historical Account of those Parishes in the County of Middlesex which are not described in the environs of London,* 1800.
4 *Northampton Mercury,* 15 November 1794.
5 Grand Junction Lower District Committee minutes, 16 December 1793.
6 35 Geo III, c 8—5 March 1795.
7 Grand Junction General Committee minutes, 15 November 1797.
8 Legg, Edward. Fenny Stratford Bridge, *The Lockgate,* January 1968.
9 *Northampton Mercury,* 31 May 1800.
10 Grand Junction Lower District Committee minutes, 20 May 1794.
11 *Northampton Mercury,* 25 June 1796.
12 Ibid.
13 Ibid, 17 September 1796.
14 The precise opening date for the railroad is not known. It was still incomplete on 3 November 1800, but rules for its operation were approved on 21 February 1801.

Blisworth and Wolverton 3

ALTHOUGH THE Grand Junction now had a through route its line was punctuated firstly by the railroad over Blisworth hill and secondly by the temporary canal across the Great Ouse valley. It was to be over five years before either of these handicaps was overcome.

From its northern summit at Braunston, on the Northamptonshire Uplands, the canal had to cross the Nene valley and then traverse a ridge of hills before descending to cross the Great Ouse at Wolverton. The ridge was fairly narrow and impracticable to cross by locks. Water supplies were almost non-existent and the summit could only have been fed by expensive pumping. There was little alternative to tunnelling, and a site near Blisworth was chosen where the ridge was at its narrowest.

As the tunnel was going to be a major undertaking Barnes opened negotiations with prospective contractors early in June 1793, only a few weeks after the Act had been obtained. In December a contract was let to Charles Jones[1] and John Biggs to do the work jointly at £10 per yard. The tunnel was to be completed by 1 June 1797.

Work started immediately. In May 1794 Jessop reported that Biggs was progressing well with his preparatory works. He had manufactured many bricks and collected other necessary materials, but had not started excavations underground. Instead he was concentrating on draining off the water that would have made tunnelling extremely difficult.

Jessop spoke less well of Jones, who seems to have rushed into the project without much preparation. Working from three faces, he had completed 24yd of tunnel, but some of the brickwork was poor and required replacing. His brickmaking was somewhat behind and he had no equipment ready to raise spoil from the shafts. Jessop tempered his comments with the optimistic view that the ground was very suitable for tunnelling and, with good management, the work could be completed in much less time than at first thought necessary[2].

Soon afterwards Jones ran out of funds and had to be dismissed. Then in September Biggs threw in his contract, saying he could not complete it on the agreed terms[3]. The company now stepped in and took over the contractors' equipment and materials. Biggs was retained as a supervisor for a while, but in May 1795 he and several members of his family were dismissed over financial irregularities following an investigation by Homer into why the cost of the completed length of tunnel was vastly in excess of the estimate.

Thereafter responsibility for the work at Blisworth fell on

The north portal of Blisworth tunnel, completed in 1805 and rebuilt in 1902/3.

Barnes, who was asked to devote as much time as possible to it[4]. Despite this, by the beginning of 1796 the works were almost at a standstill as an uncontrollable amount of water had drowned out the tunnelling operations.

Jessop seems to have lost heart at this, for he advised Praed to abandon the tunnel and build locks over the hill instead so as not to delay completing the through route. He was asked to supply details of this plan, and Barnes was also asked for his views.

Both presented their reports in February. Jessop recommended against the tunnel, but Barnes considered one not only feasible but much more satisfactory, although he suggested an amended course. With this conflict outside advice was sought from engineers John Rennie and Robert Whitworth.

Their report, delivered in May, recommended building a tunnel but on a different alignment. The new route was to start slightly to the east of the original line at the north end, to cross, and then to finish about 130yd west of the original at the south end.

Barnes, vindicated, offered to complete the 3,000yd-long tunnel by June 1799 at no more than £16 per yard and under a £5,000 penalty, providing the company would finance him. His offer was accepted, and in June 1796 he was put in complete charge at Blisworth[5]. In September he was authorised to proceed with the tunnelling and the deep cuttings at both ends, but was told to restrict his expenses to £7,000 up to 1 May 1797.

With this limitation the work proceeded only slowly. In March 1797, in an effort to speed progress up elsewhere on the canal, the company decided that all work on the tunnel should stop. Barnes was told to do only such works as were necessary to preserve what had already been completed.

Barnes, however, pleaded to keep work going on the heading, a small pilot tunnel cut beneath the main tunnel to drain away the unwanted water in the hill. He realised that the first attempt had failed mainly because of the lack of such a precaution and that the main tunnel was totally dependent on the heading being completed. His pleas were heeded, but only a skeleton staff of six men was kept on for this purpose. At Stoke Bruerne the heading extended some way beyond the workings to where the land fell and the water could drain away naturally.

Meanwhile in November 1796 Barnes had been told to plan a road over the top of Blisworth hill to follow the course of the intended tunnel, and £250 was spent constructing this in August 1797. It ran for three and a half miles from Pickfords' wharf in Blisworth, across the Northampton to Towcester turnpike, and then over the hill to join the Northampton to Old Stratford turnpike. Once finished, tolls were charged for its use.

The road created an important outlet from the canal's temporary terminus at Blisworth. Many boats were already trading to the village, where wharves and warehouses and two new inns had been established by the canal, several thousand tons of coal were in store,

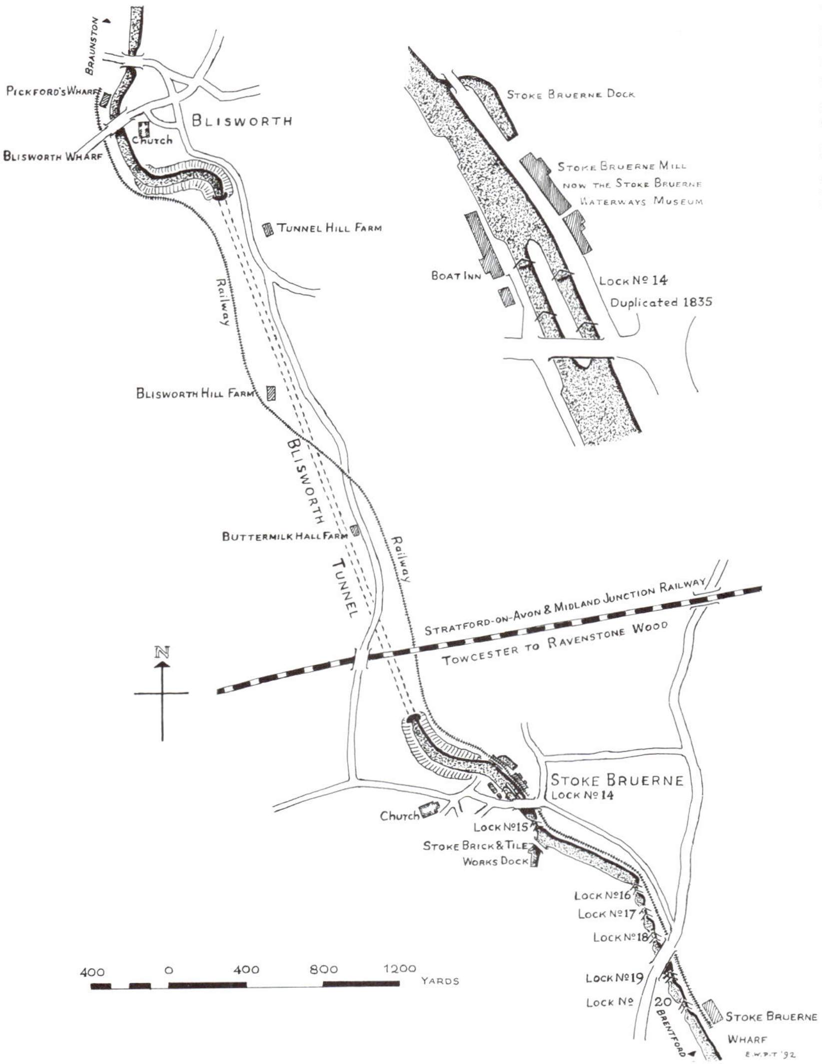

The course of the railroad laid down as a temporary measure can be seen on this map of the Blisworth and Stoke Bruerne area. The enlarged section shows Stoke Bruerne, now the site of the British Waterways Museum.

and a boatyard had been set up where craft were being built[6].

In November 1797 Barnes reported good progress on the heading. The amount of water encountered was such that driving the main tunnel would have been impossible without it. Once it was completed Barnes was confident that excavation of the tunnel could then take place along its whole length simultaneously.

By 1798, when work on the rest of the canal was proceeding well, the company began to realise that the withdrawal of the main labour force from Blisworth had been a short-sighted move. In August a proposal for Benjamin Outram, a tramroad engineer, to view the whole length of the canal was rejected, but in March 1799

Jessop proposed that a cast-iron railroad should be built from Blisworth to Cosgrove. He was asked to make a survey with Barnes.

Both engineers reported in the following month and the railroad was approved in principle at the company's general meeting in June. It seems the toll road was not very satisfactory, particularly for the heavier types of goods.

The original intention was for the line to continue beyond Stoke Bruerne by following the turnpike road to Old Stratford. Here there would be a connection with the Old Stratford branch from Cosgrove and the southern part of the canal. Later it was decided to keep the railroad as short as possible, and Barnes started work on the canal from Stoke Bruerne to Cosgrove in June 1799.

In July 1799 a sub-committee was set up under Praed to plan for the railroad or to find a more suitable way of conveying goods over the hill until the tunnel was complete. The members, accompanied by Jessop and Barnes, attended a series of tests on a colliery railway at Measham owned by Joseph Wilkes, a member of the Grand Junction's committee and a leading supporter of the railroad. The sub-committee was impressed by what it saw. After hearing from Jessop that the tunnel would take from two to three years to finish it adopted the railroad plan for Blisworth.

An immediate problem arose in that Jessop and Barnes, when planning the line, wanted to deviate more than the permitted 100yd from the Parliamentary line in places to ease some of the gradients. Permission from the local landowner, the Duke of Grafton, was essential before work could start, and Praed was deputed to approach him.

The Duke was disinclined to be helpful, since he was upset that work on the tunnel was being deferred and concerned about the effect the railroad would have on his tenants. He also wanted precautions taken to prevent those employed from becoming a nuisance in neighbouring villages. Praed tried to reassure him, but without success.

In October the company, worried that protracted negotiations would delay the project, decided to amend the line of the railroad so as to follow the course of the canal as set out in the Act. Outram had been called in and he and Barnes considered the revised line practicable; some of the gradients would be steeper but the overall length of the line would be shorter.

Outram then produced a specification. An 8yd width of well-drained land would be needed, of which the centre 6yd would be covered with a 1ft depth of stones or other hard materials. Stone block sleepers, obtainable from local quarries, were to be used. Two holes were to be drilled in each and plugged with wood. Cast-iron rails, each 1yd long and weighing 37lb, were to be fastened to the sleepers by large nails driven into the wooden plugs. The waggons would carry from one and a half to two tons.

His estimate for the double track railroad was £9,750. This included £8,525 for the 6,000yd line from Blisworth to just south of the Old Stratford turnpike at Stoke Bruerne and £1,225 for eighty waggons, four cranes and other incidentals.

The largest single item in the estimate was £5,470 for 405 tons of cast-iron rails. The stone blocks with the oak plugs and iron spikes were £910, the earthworks £665, the track bed £840, with drainage and other works £640.

Outram's estimate was accepted and a formal contract was drawn up. He had to guarantee the line for three years from the completion date, which was to be 1 August 1799. He started straightaway, but the work took much longer than expected. The exact date the railroad was opened is not known, but it was probably early in January 1801.

Soon after the opening Homer and a colleague visited similar railroads in Derbyshire to see how they were worked. Immediately afterwards a set of operating regulations was drawn up, since the need for some form of control was essential. In May a short extension was authorised at the northern end to take the line from the public wharf in Blisworth across the turnpike to Pickfords' wharf.

The Blisworth railroad would have had L-shaped rails carrying wagons with flangeless wheels.

The railroad handled all types of goods. Coal, pig iron, stone, slates, bricks, timber, lime, manufactured goods and agricultural products were all carried as through trade on the canal at the time and would have passed over the line[7]. Carriage of the goods was entirely in the hands of private contractors.

The rails used were L-shaped and were fixed to the sleepers with the horizontal flange facing outwards. The waggons had flangeless wheels which ran along the horizontal flange and were kept in position by the vertical flange[8]. From measurements taken the gauge was probably 4ft 2in, one favoured by Outram. A benefit of non-flanged wheels was that ordinary waggons could be used providing their gauge was correct. Likewise elaborate sidings were unnecessary at the two termini as the waggons could simply be run off the track and parked on a convenient piece of ground. The waggons were probably hauled up the steeper parts of the line in twos or threes and then coupled together to form trains of about a dozen behind a couple of horses.

During all this time small-scale work on the heading was kept going. It was not until November, when the railroad had been operating for nearly a year, that the company advertised for tenders to build the tunnel. Jessop and Barnes were asked to draw up proper plans and a specification, and to give an idea of the contract prices the company could expect. At the same time an assessment was made of the quantities of bricks and other materials that were already on site.

By January 1802 some preparatory work had been resumed on brickmaking at both Blisworth and Stoke Bruerne, as a partial halt was ordered until Barnes could decide what work should be let out to contract and what the company should do. Some men were retained to continue digging clay for brickmaking and another five carried on with excavating the deep cutting.

Jessop's plans were ready in May. The tunnel was to be 18ft from the crown of the arch to the bottom of the canal and 16½ft wide. The side walls and the crown were to be 17in, or two bricks, thick, and the inverted arch 13in, or one and a half bricks thick. Jessop thought the work should take two years.

He also recommended that wooden slide rails should be fitted on both sides of the tunnel 6in below water level to guide boats through, particularly those with high loads. Wooden chocks were to be fixed to the top of the rails at 9ft intervals for pushing against to propel boats through more easily than pushing against the brickwork. The cost of the rails was estimated at £1,000.

Barnes, meanwhile, had been valuing that work performed by the company which he considered should be taken over, and paid for, by the contractor. The largest item was 600,000 bricks valued at £9,450. The list included work on sinking some of the shafts, providing carts, timber, centering and hoists, with three brick kilns at Blisworth and another at Stoke. The total was almost £2,100. Barnes also recommended a further three kilns at Stoke should be

A busy scene at the northern end of Blisworth tunnel in the late 1920s. A train of four boats has just left the tunnel, whilst a southbound train is being formed up ready for the tunnel tug's next trip. *National Waterways Museum*

taken over from Joseph Ludlam, who had been making bricks for the company.

The most important work carried out by the company was the driving of the main heading, and certain cross headings, to drain away the water. At Blisworth 610yd of heading had been built at £1.80 per yard. At Stoke 1,608yd of heading and 404yd of cross heading had cost £2.12 per yard. Barnes felt this work should not be brought into the contract, but the cost of any cross headings still to be driven, and the cost of clearing away the water, must be borne by the contractor.

Jessop's and Barnes' plans were approved and the company advertised again for offers to complete the tunnel. Shortly after the 5 June closing date the tender submitted jointly by Jonathon Woodhouse, a civil engineer from Ashby de la Zouch, his brother John, a civil engineer from Chilvers Coton, George Tissington of Ashby, and his brother Anthony, a canal tunneller also from Ashby, was accepted. Joseph Wilkes and William Hill of Measham agreed to join with the contractors in a £10,000 bond for the due completion of the works in thirty months.

The contract specified that the tunnel was to be built "in a good substantial and workmanlike manner" for £15.65 per yard, including the oak rail recommended by Jessop. The tunnel was also to be maintained for one year from completion.

The company agreed to advance £3,000 immediately and to make stage payments as the work progressed against certificates from Barnes. It imposed a £3,000 retention until the work was finished and a further £1,000 to the end of the maintenance period. If the contract was completed in twenty-seven months a £1,000 bonus was to be paid.

There was some delay before the contractors began work, and then they were hampered by bad weather. In September 1802 Barnes suggested they should devote their resources and limited supply of bricks to sinking all the shafts to the level of the tunnel and completing the first length of tunnel on either side of each shaft. This was the most difficult and expensive part of the work. By doing this Barnes felt the final completion date could be brought forward by six months, as it would open up working faces all along the line. Barnes calculated the cost of this work at £7,050 over an eight-month period. In addition fifty men would be needed in the deep cuttings at either end of the tunnel costing £420 per month, with a further £100 per month for horses, hoists and other items.

The contractors' stage payments were based on the actual length of tunnel completed. As this would be small under Barnes' revised plan they would have been at an obvious disadvantage. To offset this they were to be allowed £1,000 per month for the eight-month period. After 1 May 1803 the payment system was to revert to the agreed one against Barnes' certificates, with the contractors repaying the advance at £1,000 per month. By November 1802 no less than 134yd had been finished and work was going so well that it was expected the tunnel would be opened well ahead of schedule[9].

In December Jessop proposed a change of plan. Originally there were to have been two tunnels—the main one, 2,900yd long in a straight line under the hill, separated by a short curved cutting from the second, 100yd long, at the Blisworth end—to allow the canal to follow the natural line of the valley at Blisworth and avoid a deep cutting through ground expected to be treacherous for excavation. It also contained several springs that would make the sides of the cutting likely to slip.

Barnes had reviewed the plan and found that if the main tunnel was extended northwards the second tunnel could be dispensed with. Much of the soil, originally thought to be difficult, had been found to be good clay and was being used for brickmaking. Then a heading had been driven into the cutting and succeeded in tapping the springs. Although a 48ft deep cutting would still be needed, Jessop agreed with Barnes and suggested further measures to counteract any slips occurring.

The contractors were now asked to undertake this additional length on the same terms as the main tunnel. At first they refused and sought a higher price, but when the company threatened to give the work to another firm which had offered a lower price, the contractors agreed the company's terms.

By June 1803 over 640yd of tunnel had been finished[10], and progress was speeding up, 77yd being completed in August, and 145yd in December. There was an accident early in October when two workmen were killed, as they were being hauled up one of the shafts in a basket; the basket slipped off the hook and fell back down the shaft, a drop of 20yd. One man died instantly and the other survived only a few hours[11].

Despite the good progress, the company was becoming increasingly concerned by the sums of money advanced to the contractors to help them provide materials. As the materials were used Barnes made appropriate adjustment in his certificates of work completed. Towards the end of 1803 Barnes was told to stop allowing for materials and to start making regular deductions from payments so that the amounts advanced would be repaid by the time the tunnel was expected to be finished.

Early in February 1804 the company refused to honour a bill of exchange for over £1,930 drawn by the contractors, supposedly for work on the main heading, for which the company was responsible under the contract, whilst the contractors were liable for the cross headings. Insufficient details had been provided of the work represented by the bill.

In the ensuing investigations the contractors disclosed losses of up to £13,000. They asked to be released from the contract as they could not complete it on the agreed terms. Part of the loss was from quite unforeseen difficulties, and the company paid them £2,040 as compensation, but most was due to their own bad management.

The company realised that to try and enforce the contract

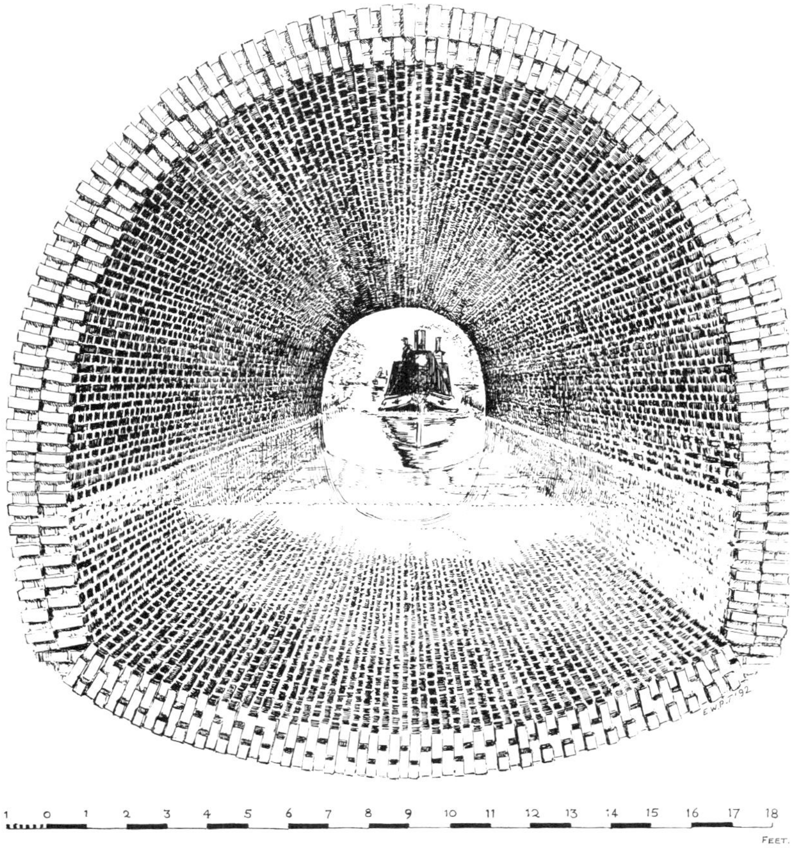

A cross-section inside Blisworth tunnel, drawn by Edward Paget-Tomlinson.

could involve a lengthy legal battle, during which time the works would be at a standstill, whilst the £10,000 performance bond would not compensate for any lengthy delay. Instead the contract was dissolved, with the contractors closing off their books on 21 March.

Barnes and two independent assessors then valued the work done, the equipment and stock of materials. They put the length of tunnel completed at 1,618yd, and commented on the high quality of the work[12].

There was a last-minute effort by Jonathon Woodhouse and Edward Mammot, one of the valuers, to take over the contract but the company declined their offer and placed Barnes in charge of the tunnel works once again. At the same time Caleb Maullin was appointed superintendent of trade at Blisworth, primarily to oversee the railroad but also to assist Barnes and to be responsible for the purchase of materials.

Barnes now engaged small groups of workmen to carry out short lengths of the tunnelling work. In April two offers to

Opposite page: An imaginative reconstruction of the railway over Blisworth Hill, opened in 1801 to link the two lengths of canal before the tunnel was opened. This wharf at Stoke Bruerne remained when in 1805 the tunnel was completed and the railway was lifted.

complete the entire work by the end of February 1805 were received, but the company refused to let any large new contract. In May John Woodhouse, one of the former contractors' partners, was taken on to assist Barnes in superintending the tunnel workmen. In August Maullin was dismissed for failing to keep proper accounts and Henry Provis was appointed in his place.

Following the takeover the previous rate of progress was not maintained. Most of the stock of bricks had been used up as the weather had been unsuitable for making new bricks. Instead Barnes concentrated on completing the heading under the tunnel and sinking an extra shaft to give two more working faces. In July some alarm was caused when a 20yd section of tunnel collapsed; the fallen part was quickly replaced and the brickwork was strengthened as a precaution. In the same month Barnes was told to complete the tunnel portals.

Thereafter the work proceeded uneventfully, with Praed and other members of the committee keeping a close watch on progress. By September only 980yd remained to be driven, and four new shafts had been sunk[13]. The shortage of bricks had also been overcome. Two months later 2,500yd had been completed, leaving only 570yd outstanding[14]. More would have been achieved had wet weather not slowed the work somewhat. By the end of January only 88yd remained, and the tunnel was finally finished on Monday, 25 February 1805[15].

Meanwhile early in February 1804 Henry Provis started preparations to build the locks at Stoke Bruerne that would take the canal down to the end of the railroad and to the stretch of canal already in use, and in May excavation for the foundations of the first two locks was started. They were all built by contract. Initially William Jackson was working on all seven, but his progress was too slow and in July he agreed to confine his efforts to just three. A contractor from Buckingham agreed to build three more and to take over the materials Jackson had collected for them. By September the work on all the locks was well advanced and they were completed in time for the opening of the tunnel.

As soon as the tunnel was finished Thomas Telford, a well-known canal engineer, was invited to inspect it. He did so in March, and reported it perfectly straight and the materials and workmanship excellent[16].

Overleaf: A map showing the Grand Junction Canal and its associated waterways. The Grand Junction and its branches are shown in red and the Leicester line in brown; the connection to Birmingham by way of the Warwick and Birmingham and Warwick and Napton canals is in blue.

The company decided to hold a formal opening ceremony on Monday, 25 March 1805. It was a major event. The first boat through, a packet boat hired for the official party, joined others assembled at the north end to form a grand procession. The tunnel was lit by flambeaux and other lights for the occasion. The official party took an hour and two minutes to pass through to the south end, where 5,000 people were gathered to greet them[17]. A dinner for 100 people was provided at the Bull Inn, Stony Stratford, costing half a guinea each.

The cost of the second, successful, attempt was at least £90,000.

R. DEE
CHESTER
ELLESMERE PORT
CHESTER CANAL
MIDDLEWICH
MIDDLEWICH BRANCH
RUNCORN
MERSEY
MANCHESTER
MACCLESFIELD C.
HARECASTLE TUNNEL
THE POTTERIES
NANTWICH
NEWCASTLE UNDER LYME
CALDON CANAL
LEEK
FROGHALL
UTTOXETER BRANCH
UTTOXETER
ELLESMERE CANAL
WHITCHURCH
PREES BRANCH
ELLESMERE, LLANGOLLEN WELSHPOOL, NEWTOWN
BIRMINGHAM & LIVERPOOL JUNCTION CANAL
TRENT & MERSEY CANAL
CROMFORD
CROMFORD CANAL
PINXTON
LANGLEY MILL
NOTTINGHAM CANAL
LITTLE EATON BRANCH
NUTBROOK CANAL
DERBY CANAL
EREWASH CANAL
DERBY
DERBY CANAL
LOUGHBOROUGH NAVIGATION
KEGWORTH
R. TRENT
HORNINGLOW
BURTON ON TRENT
GREAT HAYWOOD JUNCTION
STAFFORD
R. SOW
NORBURY
NEWPORT BRANCH
NEWPORT
POOL QUAY (WELSHPOOL)
SHREWSBURY
SHREWSBURY CANAL
R. TERN
DONNINGTON WOOD CANAL
COALBROOKDALE
SHROPSHIRE CANAL
IRONBRIDGE
STAFFORDSHIRE & WORCESTERSHIRE CANAL
ARMITAGE TUNNEL
FRADLEY
LICHFIELD
HUDDLESFORD JC.
WHITTINGTON BROOK JC.
TAMWORTH
WYRLEY & ESSINGTON CANAL
ASHBY DE LA ZOUCH
MOIRA
LOUGHBOROUGH
FOREST LINE
MOUNTSORREL
LEICESTER NAVIGATION
SNARESTONE TUNNEL
ASHBY CANAL
POLESWORTH
ATHERSTONE
FAZELEY
BIRMINGHAM & FAZELEY C.
COVENTRY C.
NUNEATON
GRIFF BR.
ARBURY CANALS
BEDWORTH
MARSTON JUNCTION
HAWKESBURY JUNCTION
WYKEN BRANCHES
COVENTRY
OXFORD CANAL
NEWBOLD TUNNEL
RUGBY
BLABY
FOXTON LOCKS
KILWORTH
AUTHERLEY JC.
ALDERSLEY JC.
WOLVERHAMPTON
TAME VALLEY CANAL
DUDLEY
BIRMINGHAM CANAL NAV.
SALFORD JUNC.
BIRMINGHAM & WARWICK JC. CANAL
BORDESLEY JUNC.
WARWICK & BIRMINGHAM CANAL
BRIDGNORTH
RIVER SEVERN
R. STOUR
DUDLEY C.
STOURBRIDGE CANAL
DUDLEY EXTENSION C.
KINGS NORTON
KINGS NORTON TUNNEL
KNOWLE
KIDDERMINSTER
WEST HILL TUNNEL
BEWDLEY
STRATFORD UPON AVON CANAL
LAPWORTH
KINGSWOOD JUNC.
SHREWLEY TUNNEL
HATTON LOCKS (21)
WARWICK & NAPTON CANAL
WARWICK
KAYE'S BR.
OXFORD CANAL
NAPTON
BRAUNSTON
BRAUNSTON TUNNEL
GRAND UNION CANAL
SOUTHAM
STOURPORT
TARDEBIGGE
WORCESTER & BIRMINGHAM CANAL
LEOMINSTER CANAL
SOUTHNET TUNNEL
LEOMINSTER
DICK BROOK
DROITWICH JUNC: C.
R. SALWARPE
DROITWICH CANAL
WORCESTER
DIGLIS BASIN
R. TEME
STRATFORD UPON AVON
RIVER AVON
PERSHORE
CROPTHORNE
EVESHAM
BANBURY
OXFORD CANAL
RIVER LUGG
HEREFORD
HEREFORDSHIRE & GLOUCESTERSHIRE CANAL
OXENHALL TUNNEL
TEWKESBURY
COOMBE HILL CANAL
NEWENT
RIVER WYE
GLOUCESTER DOCKS
MONMOUTH
GLOUCESTER & BERKELEY CANAL
FRAMILODE
SAUL
STROUD
STROUDWATER CANAL
CHALFORD
SAPPERTON TUNNEL
DUKE'S CUT
ISIS LOCK
OXFORD
R. THAMES
LECHLADE
INGLESHAM
CIRENCESTER
LYDNEY CANAL
SHARPNESS DOCKS
THAMES & SEVERN CANAL
ABINGDON
CHEPSTOW
NORTH WILTS CANAL
WILTS & BERKS CANAL
WANTAGE
SWINDON
BRISTOL CHANNEL
AVONMOUTH
BRISTOL
CHIPPENHAM
CALNE
KENNET & AVON CANAL
NEWBURY
BATH
AVON NAVIGATION
SOMERSETSHIRE COAL C.
RADSTOCK BR.
SEMINGTON
DEVIZES
HONEYSTREET
CROFTON
BRUCE TUNNEL

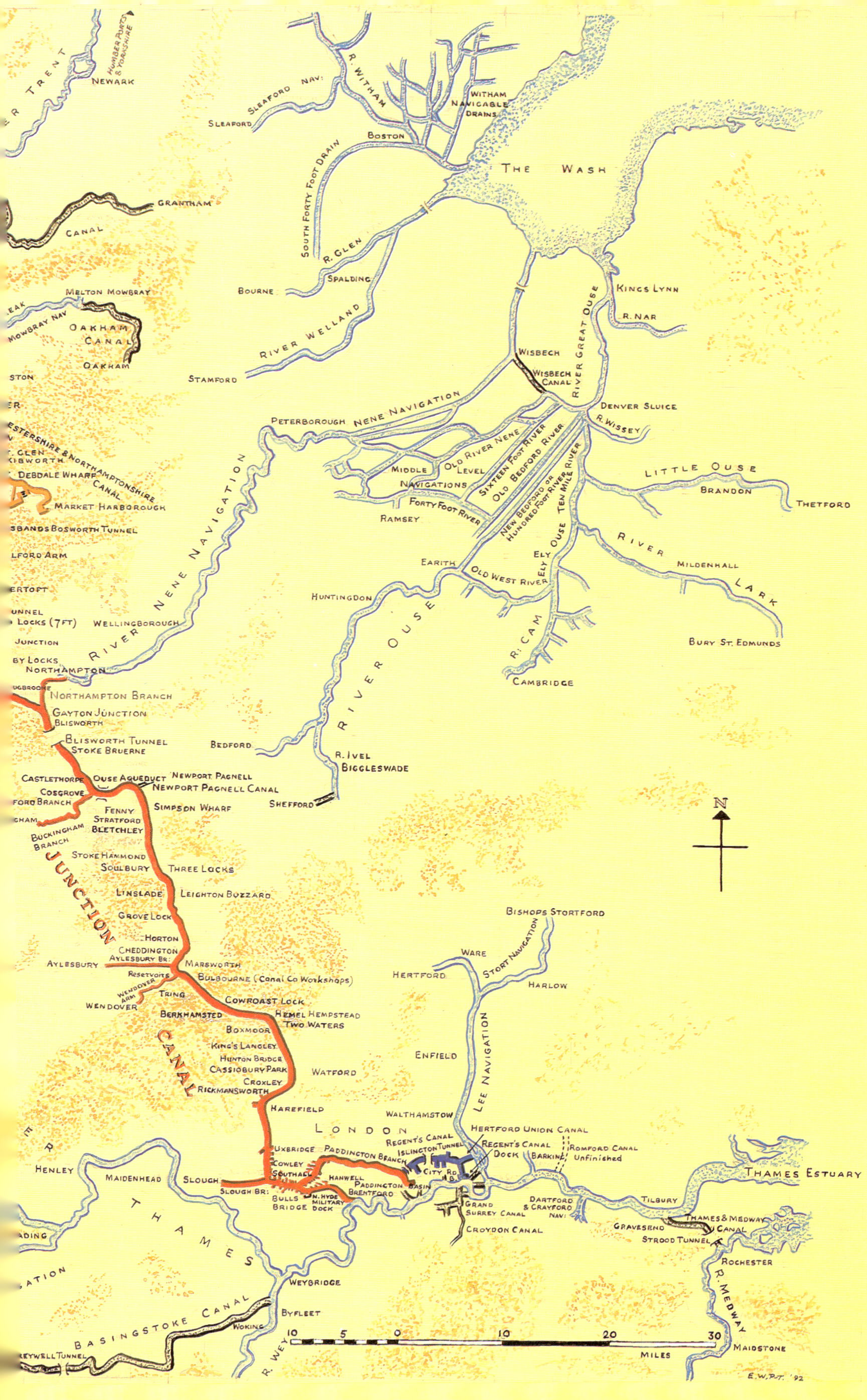
HUMBER PORTS & YORKSHIRE
NEWARK
R. TRENT
SLEAFORD NAV:
SLEAFORD
R. WITHAM
WITHAM NAVIGABLE DRAINS
BOSTON
THE WASH
SOUTH FORTY FOOT DRAIN
GRANTHAM
CANAL
R. GLEN
SPALDING
BOURNE
KINGS LYNN
R. NAR
MELTON MOWBRAY
MOWBRAY NAV
OAKHAM CANAL
OAKHAM
RIVER WELLAND
STAMFORD
WISBECH
WISBECH CANAL
RIVER GREAT OUSE
DENVER SLUICE
R. WISSEY
PETERBOROUGH
NENE NAVIGATION
MIDDLE LEVEL NAVIGATIONS
OLD RIVER NENE
SIXTEEN FOOT RIVER
OLD BEDFORD RIVER
NEW BEDFORD OR HUNDRED FOOT RIVER
TEN MILE RIVER
LITTLE OUSE
BRANDON
THETFORD
LEICESTERSHIRE & NORTHAMPTONSHIRE CANAL
R. GLEN
KIBWORTH
DEBDALE WHARF
MARKET HARBOROUGH
HUSBANDS BOSWORTH TUNNEL
RAMSEY
FORTY FOOT RIVER
RIVER NENE NAVIGATION
ELY OUSE
ELY
RIVER LARK
MILDENHALL
EARITH
OLD WEST RIVER
HUNTINGDON
R. CAM
BURY ST. EDMUNDS
WELLINGBOROUGH
LOCKS (7FT)
JUNCTION
NORTHAMPTON
CAMBRIDGE
RIVER OUSE
NORTHAMPTON BRANCH
GAYTON JUNCTION
BLISWORTH
BLISWORTH TUNNEL
STOKE BRUERNE
BEDFORD
R. IVEL
BIGGLESWADE
CASTLETHORPE
OUSE AQUEDUCT
NEWPORT PAGNELL
NEWPORT PAGNELL CANAL
COSGROVE
SHEFFORD
SIMPSON WHARF
FENNY STRATFORD
BLETCHLEY
BUCKINGHAM BRANCH
N
STOKE HAMMOND
SOULBURY
THREE LOCKS
LINSLADE
LEIGHTON BUZZARD
GROVE LOCK
HORTON
CHEDDINGTON
AYLESBURY BR:
AYLESBURY
MARSWORTH
BISHOPS STORTFORD
STORT NAVIGATION
WARE
HERTFORD
HARLOW
Reservoirs
WENDOVER ARM
WENDOVER
TRING
BULBOURNE (Canal Co Workshops)
COWROAST LOCK
BERKHAMSTED
HEMEL HEMPSTEAD
TWO WATERS
BOXMOOR
KING'S LANGLEY
HUNTON BRIDGE
CASSIOBURY PARK
CROXLEY
RICKMANSWORTH
GRAND JUNCTION CANAL
ENFIELD
WATFORD
LEE NAVIGATION
HAREFIELD
WALTHAMSTOW
LONDON
HERTFORD UNION CANAL
REGENT'S CANAL
ISLINGTON TUNNEL
REGENT'S CANAL DOCK
ROMFORD CANAL Unfinished
BARKING
UXBRIDGE
PADDINGTON BRANCH
COWLEY
SOUTHALL
CITY RD.
HENLEY
MAIDENHEAD
SLOUGH
SLOUGH BR:
HANWELL
PADDINGTON BASIN
BRENTFORD
BULLS BRIDGE
N. HYDE MILITARY DOCK
GRAND SURREY CANAL
DARTFORD & CRAYFORD NAV:
TILBURY
THAMES ESTUARY
THAMES & MEDWAY CANAL
GRAVESEND
STROOD TUNNEL
CROYDON CANAL
THAMES
ROCHESTER
R. MEDWAY
WEYBRIDGE
BYFLEET
BASINGSTOKE CANAL
WOKING
GREYWELL TUNNEL
R. WEY
10 5 0 10 20 30
MILES
MAIDSTONE
E.W.P.T. '92

SAMUEL BARLOW COAL Co Ltd
BIRMINGHAM No 69
WARWICK 2

The tunnel was 3,075yd 2ft in length and 16½ft wide, allowing two narrow boats to pass. From water level it was 11ft 3in high to the crown of the arch, with the water 5ft 9in deep in the centre.

Cast-iron markers were placed every 100yd along the west wall, numbered from the north end. Jessop's wooden slide rails did not survive long and were eventually removed, probably because they reduced the width of the tunnel too much. They also offered a somewhat cumbersome way of working boats compared to legging (see Chapter 7), which became standard. As the tunnel is perfectly straight it is possible to see through it when it is not obscured by fumes.

Construction of the tunnel was a major undertaking for the engineers and contractors of the day. The first step in planning the work was to set out a straight line over the ground above the line of the tunnel. A longitudinal section was then taken showing the rises and falls in that straight line.

This section was then plotted on paper to a suitable scale and a straight line drawn in to indicate the level of water in the tunnel underneath the ground contour. The longitudinal section was then divided into an equal number of parts and, by scaling, the depth of the canal below ground level could be obtained at every point.

Translating this to the actual ground gave the depth to which the shafts at each of the points should be sunk. Horizontal excavation along the line of the tunnel was then started in both directions from the foot of each shaft. In due course all these excavations met, resulting in a straight and level tunnel.

To set out a straight line one and three quarter miles over Blisworth hill by sighting required one point high enough for the engineers to be able to see the full length of the line. It is no coincidence that the tunnel lines exactly on the tower of St Mary's Church at Stoke Bruerne, which stands on a high piece of ground; the view from the tower is extensive, making it an ideal sighting point.

Opposite page: Although the original three-arched brick aqueduct over the Great Ouse at Wolverton collapsed in February 1808 because of its poor construction, the cast-iron replacement completed in 1811 has remained firm. The trough is constructed from segmental flanged plates supported by a central stone pier and reinforced by two sets of arched members. London-bound with coal is the Samuel Barlow butty *Warwick,* built in 1935 by Walkers' of Rickmansworth as the *Ethiopia* for the Grand Union Canal Carrying Co Ltd and renamed when acquired by Barlow's in 1945. She was for long paired with the *Cairo* under George Smith and his wife Sonia, who later became Sonia Rolt and has written the foreword to this book.

Setting out Blisworth tunnel.

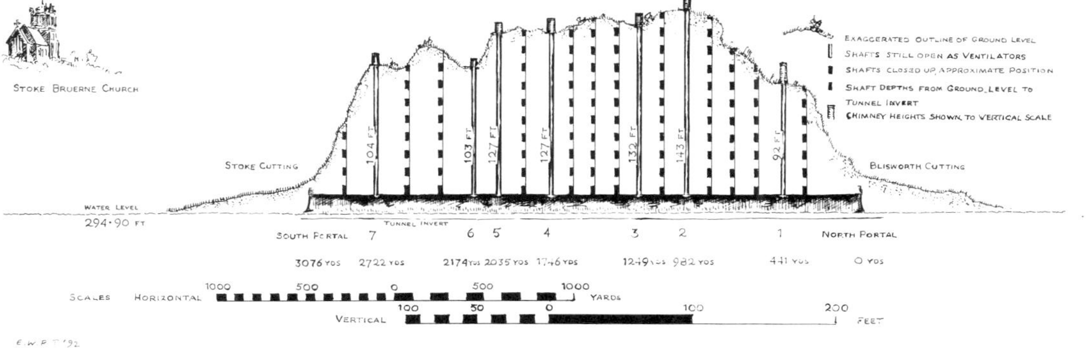

Right: The cross stank in use in Blisworth tunnel, enabling a short section of the tunnel to be dammed off to allow repairs to the invert without having to drain the entire tunnel. *National Waterways Museum*

Opposite page, top: At least nine pairs of boats waiting below Stoke Bruerne locks in April 1907 during a stoppage. The Easter and Whitsun bank holidays were the usual times for stoppages, as many canal-served wharves and factories would be closed. *National Waterways Museum*

Opposite page, bottom: A pair of boats prepare to leave Stoke Bruerne top lock, whilst several other boats are tied up above the lock. The tall chimney marks Stoke Bruerne mill, which was served by a short dock. *Michael Hooper*

In all nineteen shafts were sunk, but most were filled in once construction was complete. Initially four were retained to provide ventilation, recognisable from the brick shafts on the ground above. These are like chimneys and were built to create a draught as well as to prevent debris falling down into the tunnel below. The sites of the filled-in shafts can still be seen along the line of the tunnel as mounds some 100ft in diameter. These mounds were formed as earth was drawn up from the excavations below by horse-operated hoists, or gins as they were known.

With the opening of the tunnel the railroad and transhipment facilities were redundant. The company was able to use the expensive cast-iron rails and many of the stone sleepers on the Northampton railroad (see Chapter 4), but the carriers who operated over the line sold off most of their rolling stock. The company also sold much of the equipment used at the tunnel, disposing of 350 tons of cast and wrought iron, five mills for grinding brick clay, fourteen pit gins with their ropes and pulleys, and many other items.

It is hardly surprising that a tunnel that was such a struggle to build should prove to be a constant headache to successive maintenance engineers. The ground through which it was driven—clay, rock, limestone and ironstone—is extremely liable to move, particularly during wet weather. These movements have caused serious distortions to the arch, walls and invert. In contrast, Braunston tunnel, only eighteen miles to the north, has been comparatively trouble free and required little attention.

Wolverton Aqueduct and Embankment

The second handicap to the company's through route was the temporary canal across the Great Ouse valley.

Once the general line of the canal had been settled and the two summits, at Tring and Braunston, had been fixed there was no alternative to a series of locks into and out of the Great Ouse valley, without an unacceptably long detour. In the original plans the canal was shown as crossing the river on the level between Old Wolverton and Cosgrove, the drop of 190ft from Tring being achieved by thirty-one locks, and the rise of 160ft to Braunston by another twenty-eight locks.

The Great Ouse has always been prone to flooding, and canal traffic could have been seriously delayed at the crossing point. In December 1799 Barnes suggested an aqueduct over the river. This would be a major improvement, and he was told to prepare plans and consult with the landowners involved[18]. His proposals were soon agreed in principle, the cost being put at £25,000. Four locks on each side of the valley would be eliminated. The decision was taken just in time to alter the plans for the canal between Stoke Bruerne and Little Woolstone to follow the contour and eliminate the locks.

In April 1800, when the company was concentrating on completing its through route, Jessop estimated the aqueduct would take two years to build. Meanwhile he recommended a temporary canal across the Great Ouse valley on the southern side of the proposed embankment. Eight locks would be needed to overcome the 34ft drop from Old Wolverton to the river and the climb out again to Cosgrove.

Jessop envisaged less substantial locks than permanent structures. The inverts could be of timber, saving 80,000 bricks as well as the time and expense laying them. It was intended that the gates and other fittings would be transferred later to the Stoke Bruerne flight, where construction had been deferred until Blisworth tunnel was complete.

At the time of Jessop's report work was well advanced on the canal from Fenny Stratford to Wolverton and from the end of the railroad at Stoke Bruerne to Cosgrove. Construction of the temporary canal across the valley was started immediately, and was completed in September.

Nothing further was done towards the aqueduct until May 1802. Jessop was then asked to prepare a plan, section and estimate. He produced these in June and recommended an early start[19]. At the time there was much local trade across the Ouse valley, which needed a large amount of water from the Tring summit, where it could be ill afforded. To help cope with the situation whilst the aqueduct was under construction a forty-acre reservoir was built at Wilstone on the summit near Tring.

It was company policy at the time to undertake large works by contract, so the Wolverton scheme was put out to tender. At the end of November an offer submitted jointly by Thomas Harrison of Wolverton, Major Mansel, Thomas and Joseph Kitchen of Castlethorpe and William Oliver of Stony Stratford was accepted. Their tender was 6p for every cubic yard of earth including the puddling, and £12 for every rod of brickwork. They had to deposit £5,000 as surety. The company's solicitors then drew up a formal contract and the contractors began preparations. Later the contractors agreed to widen a narrow cutting at Old Wolverton at 12½p per cubic yard.

In January 1803 Harrison, chief partner in the Wolverton Valley Company, as the syndicate was known, agreed the final contract terms. He had been Earl Spencer's principal agent. The contract was to be completed within two years.

Work started immediately, and in August 1803 the canal company appointed Henry Provis to oversee operations at Wolverton. It also assisted by allowing the contractors to use the toll house and store at Cosgrove, by hiring out railway lines bought for an intended rail-road to Aylesbury, by providing floats to ferry materials, and by contributing half the cost of a pumping engine to drain a section of the works.

Early in 1804 it was apparent the contract was not proceeding precisely to plan. Provis reported that the contractors intended making several additional charges for the work. They were asked for full details of these extras whilst Provis and Benjamin Bevan, an engineer from Leighton Buzzard, were asked to prepare fresh estimates taking the extra charges into account.

In April the figures prepared by Provis and Barnes were approved as the basis on which to negotiate with the contractors. An agreement was reached in May and the solicitors amended the contract accordingly, incorporating in it the agreement to widen the cutting at Old Wolverton.

By June two-thirds of the work on the embankment had been finished, one-third since the previous November. Half the work on the aqueduct had been done and it was expected to be open well in time for the opening of Blisworth tunnel.

In October Provis challenged the way the arches were being

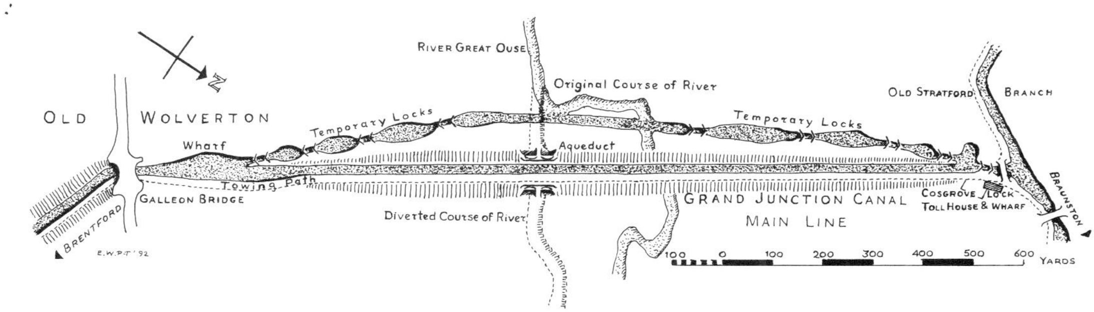

The original canal across the Great Ouse valley at Wolverton was later by-passed by a high-level canal just to the north, eliminating eight locks.

built. Charles Handley, who had engineered the Warwick & Napton Canal, was asked to inspect the work. Provis was overruled as Jessop had ordered the arches to be built in that way, and Handley agreed with him.

In April 1805 there was an alarmist report that the aqueduct had collapsed[20]. The committee met at once and issued a denial saying that Telford had just examined the works and found everything satisfactory[21].

By the middle of August a restricted channel was open over the aqueduct and embankment. As the water supply situation was again critical the company's treasurers suggested that the embankment should be opened for trade until the beginning of November. They undertook responsibility for the work in return for half of the additional revenue expected to be generated. The contractors agreed and the aqueduct was opened for traffic on Monday, 26 August 1805[22].

No plans of the aqueduct have survived, but it was a three-arched brick structure. The arches were semicircular and ran under the full width of the embankment at river level, rather like enlarged culverts. They were constructed on dry land and the river was then diverted from its original course nearer to Cosgrove to pass through them. For many years gradual subsidence took place in the embankment over the old river crossing and it had to be consolidated by timber piles driven into the slope.

The aqueduct remained in use beyond the start of November as originally proposed, but in January 1806 a section of the embankment near Cosgrove collapsed[23], almost certainly at the point of the old river crossing. This put an abrupt end to the use of the new route. A sub-committee was set up under Praed's leadership to investigate the position and decide on a suitable course of action. Soon afterwards Rennie was called in to report on the state of the works.

The problems were attributed to bad workmanship. A letter was sent to Harrison insisting on the company's right to have the works repaired and holding him liable for damages for the loss of trade suffered. He rejected the claims and a lengthy dispute ensued. Harrison claimed the company was responsible to repair the damage as the work had been completed in accordance with the contract and the contractors were still owed money. The company claimed the works had never been properly completed and the contractors must repair the damage and compensate the company for the loss of traffic dues.

Matters were no further forward nearly a year later. Harrison was still claiming the works had been completed and, as no further action was necessary, the contractors were withdrawing from the site on 1 July.

An architect was then called in to give an independent opinion on the aqueduct. His report claimed it was in a dangerous and imperfect condition. The piers had sunk and were out of alignment, the counter arches had risen in the middle, and the structure had most certainly not been completed in accordance with the plans. At the same time Provis and the local overseer reported on the embankment. Their report indicated that the contractors had made temporary repairs to the breach. More repairs were still needed as the embankment had not been finished to the contract specification.

Harrison was then given an ultimatum. If his claims against the company had not been received by 29 September legal proceedings would be started against him and his fellow contractors to recover the monies paid to them for the work. In October Harrison at last sent in his claim, and Provis and Bevan were asked to examine it and negotiate a speedy settlement.

Before any agreement could be reached the dispute took a dramatic turn when the aqueduct collapsed on the night of Tuesday, 18 February 1808[24].

The iron trough of Wolverton aqueduct, showing the central stone pier and the brick abutments. *Alan Faulkner*

> Stony Stratford. Feb 22.
> On Friday morning last the inhabitants of this town were thrown into the utmost consternation, by information which arrived from Wolverton, that the large aqueduct arches under the immensely high embankment for carrying the new line of the Grand Junction Canal across our valley, about a mile below this town, had fallen in; and that the river Ouse was so dammed up thereby, that this town must shortly be entirely inundated to a great depth. On repairing to the spot, however, it was found, that one of these arches, which had been propped up underneath with timber, soon after the centres were removed, was still standing; and that this one arch, owing to there being no flood in the river, was able to carry off the water as fast as it came down. On examining the other two arches it appeared that about 22 yards in length of the middle part of each had fallen in, and blocked up the arches, laying the canal above in complete ruins, emptying it as far as the nearest stop-gate on each side, and exposing the remains of 500 quarters of coke and cinders, which the contractors had laid on the arches. The ends of each of the broken arches were found standing in a crippled state,—most fortunately for the public as well as the Company the old line of canal and locks across this valley are still standing, and in sufficient repair immediately to convey the barges, and prevent interruption to trade, but the loss of £400 a month, which we are told has of late been the extra tonnage received by the company for goods passing over this canal, will be lost to them during the period of rebuilding the arches and repairing the canal above them.

William Cherry, the toll clerk at Cosgrove lock, was the first person to see the disaster. At 11pm he just had time to let off some of the water before the collapse. He despatched a messenger to warn the inhabitants of Stony Stratford of the accident[25]. Luckily there was only minor flooding.

The Grand Junction committee met at once to hear that the northern arch, still standing, was in a dangerous condition. On Jessop's orders the earth over it had been removed, the men having worked all day on Sunday on this vital task. Harrison was asked for his proposals for rebuilding the aqueduct, whilst Provis was asked to report on its condition and to plan for a new structure. A record of the loss of tonnage was to be kept.

Harrison declined to rebuild the aqueduct, disclaiming all responsibility for the failure. Legal proceedings were then started against him. The rebuilding was made all the more urgent by the growing trade on the canal and the dry weather. Both were causing concern as the water supply situation in the long Fenny pound was serious.

In April 1808 Charles Martyn, a local carpenter, agreed to build a temporary wooden trunk across the breached aqueduct. This was opened for traffic on 10 June 1808[26]. Its cost, together with the repairs to the canal, was £2,500, but the company calculated it had lost another £1,400 during the three and a half months the local trade across the valley had been stopped. Several years later Provis was given a £52.50 gratuity for his work in planning and overseeing the erection of the wooden trunk.

By May the plans for a replacement aqueduct had been completed. They were for a conventional brick structure. No action was taken as it had also been suggested iron would be better, and Bevan supported this idea. He and the company secretary, Charles Harvey, were asked to make a detailed investigation.

They inspected both masonry aqueducts, such as that over the river Avon on the Warwick & Napton Canal, and iron such as Longdon-on-Tern on the Shrewsbury Canal and Chirk and Pontcysyllte on the Ellesmere Canal. From these visits Bevan drew

A detail of the ironwork of Wolverton aqueduct. *National Waterways Museum*

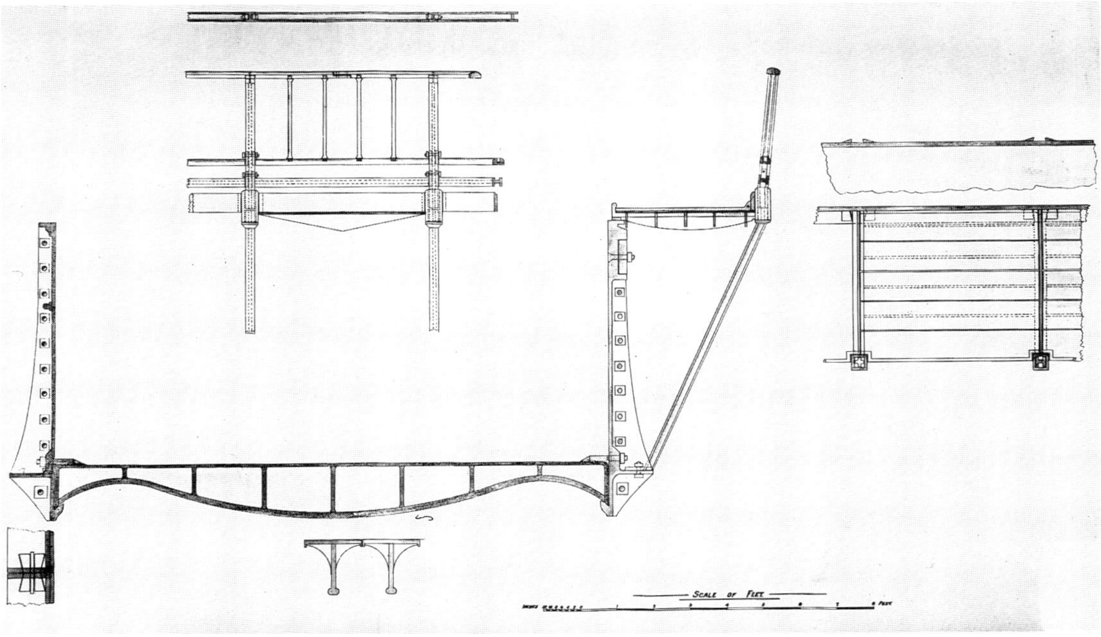

up a plan for an iron aqueduct at Wolverton, and two ironmasters, William Reynolds of Ketley Ironworks and William Hazeldine of Bridgnorth, were asked to tender for the work.

Meanwhile the action was proceeding against Harrison after his attempt to obtain a private settlement had failed. The case was heard before the Court of King's Bench on 18 July and William Bolland, a barrister, was appointed to investigate the damage and the claims of both parties[27]. At the beginning of November an award of £9,262 and costs was made against Harrison. Two of the main items were for the construction of the temporary wooden aqueduct and the making of bricks at Stoke Bruerne for the new aqueduct; these amounted to £3,223.

Harrison was now approached to settle the award. He offered to pay in instalments with £2,500 on 1 January 1809, £2,500 on 1 April, £2,500 on 1 July and the balance, with interest on the last three instalments, on 1 October 1809. The company was sympathetic and agreed, for it had emerged at the trial that the failure of the works was due to major shortcomings by those employed and not by the contractors themselves. Whilst ultimately responsible, Harrison had been unaware of the frauds being committed by people in whom he had trusted.

The first instalment was paid punctually, but Harrison died on 9 March 1809. His son, who was his father's executor, undertook to repay the debt and the remaining three instalments were paid at the end of June 1809.

In March 1809 work started collecting materials to build the pier of the new aqueduct, Bevan being asked to advise on prices and quality. At first he recommended gornall stone from Cornwall for the central pier, but later he discovered an equally suitable but cheaper stone from the Hornton quarries in Warwickshire which was used instead.

Reynolds and Hazeldine were now asked to confirm their figures, and early in April Reynolds & Co's tender was accepted in principle. They were to be required not only to supply the ironwork but to erect and maintain it for one year after completion. By May the plans had been agreed and Reynolds was asked for a detailed specification to enable a contract to be drawn up.

Reynolds attended the next committee meeting with his partner, William Anstice, and offered to contract for making and erecting the ironwork and to guarantee it for two years for £3,667.50. The company was to erect the central pier and abutments and to provide scaffolding, ropes and cranes.

This offer was accepted, and work on the new aqueduct started in the autumn with the foundation stone being laid by Bevan on 9 September 1809[28]. Soon afterwards it was decided to build the abutments in brick rather than in stone to match the pier, saving £250.

By November 1810 the aqueduct was almost complete, and it was opened for traffic on 22 January 1811 with Bevan being

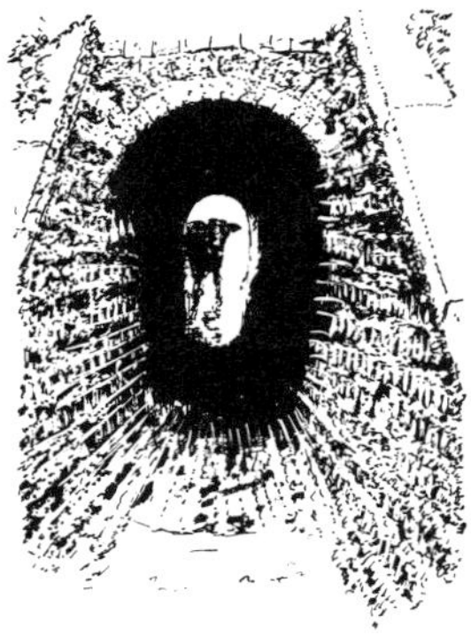

One of the cattle creeps under the canal at Wolverton aqueduct.

accompanied across in the first boat by several gentlemen from Stony Stratford[29]. The cost of the aqueduct and associated works was about £6,000. The structure has stood firmly and has never threatened to repeat the 1808 disaster.

The aqueduct is 101ft 2in long, 15ft wide, 6ft 6in from top edge to floor inside and 35ft 7in from top edge to normal river level. There are two cattle creeps under the canal, one on either side of the river. These small tunnels were driven through the embankment to give access to the far side of the canal. They are 77ft long, 6ft high and 3ft 4in wide, whilst the arch is 9ft 9in below water level.

The large lakes at the foot of the embankment on the north-east side were formed when material was excavated for the embankment. Little trace now remains of the eight temporary locks, which were dismantled when the aqueduct was finally satisfactory. The line of the foundations of the arches of the old aqueduct can still be seen in the river bed today.

NOTES TO CHAPTER THREE

1 Probably the same Charles Jones who was employed at Sapperton Tunnel. See Household, Humphrey. *The Thames & Severn Canal,* 1969.
2 Grand Junction Lower District Committee minutes, 20 May 1794.
3 Ibid, 3 September 1794.
4 Grand Junction General Assembly minutes, 1 June 1795.
5 Ibid, 7 June 1796.
6 Pitt, William. *General View of the Agriculture of the County of Northampton,* 1809.
7 Hatley, Victor. *The Blisworth Hill Railway,* 1963.
8 Ibid.
9 Report to General Assembly, 2 November 1802.
10 Report to General Assembly, 7 June 1803.
11 *Northampton Mercury,* 22 October 1803.
12 Report to General Assembly, 5 June 1804.
13 *Northampton Mercury,* 8 September 1804.
14 Report to General Assembly, 6 November 1804.
15 *Northampton Mercury,* 9 February 1805.
16 Ibid, 16 March 1805.
17 Ibid, 30 March 1805.
18 General Committee minutes, 10 December 1799.
19 Report to General Assembly, 8 June 1802.
20 *Oracle,* 13 April 1805.
21 *Northampton Mercury,* 26 April 1805.
22 Ibid, 31 August 1805.
23 Report to General Assembly, 3 June 1806.
24 *Northampton Mercury,* 27 February 1808.
25 Hassell, J. *A Tour of the Grand Junction,* 1819.
26 General Committee minutes, 14 June 1808.
27 *Northampton Mercury,* 23 July 1808.
28 Ibid, 30 September 1809.
29 Ibid, 26 January 1811.

Building the Branches 4

THE GRAND Junction was essentially a trunk waterway between London and the Midlands, but it was anxious to connect up with as many towns as possible by branch canals.

In the original Act four branches were authorised: from the Braunston summit to Daventry; from Gayton near Blisworth to Northampton and the river Nene; from Cosgrove in Northamptonshire to Old Stratford; and from Rickmansworth to Watford in Hertfordshire. The Act also authorised building feeders to supply the main canal with water.

In June 1793 Barnes was instructed to survey possible branches to Aylesbury, Buckingham, Chesham, Dunstable, Hemel Hempstead, Newport Pagnell, St Albans and Wendover. He had completed his plans by September. Three branches, to Chesham, Hemel Hempstead and Newport Pagnell, were eliminated, but the rest were referred to Jessop for his opinion.

Jessop approved the plans and a Bill was promoted to authorise all five. In December the Dunstable branch was withdrawn because of opposition from landowners. Two months later the St Albans branch was also withdrawn as opposition was expected. Here, though, it was decided to bring in a separate Bill in the next session of Parliament.

The three remaining branches, from the Old Stratford branch to Buckingham, from Marsworth to Aylesbury, and from Tring to Wendover, were authorised in the second Act[1] in March 1794.

At the end of May 1794 the company decided to revive the St

A horse crosses the bridge carrying the main canal towpath across the entrance to the dock serving Stoke Bruerne mill. The dock, constructed in 1846 by George Savage, has now been filled in. *Michael Hooper*

Albans plan and to bring in Bills for this and for a branch from Southall in Middlesex to Paddington on the outskirts of London. An Act for the Paddington line[2] was passed in April 1795 and, despite the opposition, another for the St Albans branch[3] in June. These were the last successful Acts authorising branches until 1879, when the Slough branch was sanctioned.

The Aylesbury, Buckingham, Northampton, Old Stratford, Paddington, Slough and Wendover branches were built by the company, whilst that to Newport Pagnell was built by an independent company. The branches to Daventry, Watford and St Albans, albeit authorised by Parliament, were never built. There were several other proposals for links with the Grand Junction, but none succeeded.

The first branch to be completed was the navigable feeder to Wendover, needed to supply the Tring summit with water. Negotiations were started in June 1793 to buy the Marquis of Buckingham's mill at Weston Turville and in July to buy a mill at Wendover. Meanwhile Barnes drew up plans for building the feeder.

By September the Weston Turville mill had been bought for

New Mill at Tring, beside the Wendover branch, depended heavily on water transport for many years. It is still an active mill today, although the original windmill shown here has long since been demolished. *Heygates Limited*

A pair of narrow boats tied up at Buckland on the Wendover branch, probably in 1901. They belonged to Alfred Payne, one of the last traders to use this section of the canal. *National Waterways Museum*

£400 and that at Wendover for £1,750. The water rights to a mill at Halton had also been acquired and agreement reached on the course of the feeder through that village.

Construction work was well underway by the autumn, but by then it was clear little extra expense would be incurred in making the feeder navigable. Powers to do this were contained in the 1794 Act. The work then proceeded uneventfully and the six-and-three-quarter-mile level branch was completed probably some time in 1796. It had definitely been finished by March 1797[4].

Old Stratford and Buckingham Branches.

The next branch to be built was that to Old Stratford, followed by its extension to Buckingham. In the deposited plans they were shown as running from the main line junction at Wolverton up the Great Ouse valley first to the Roman road, Watling Street, at Old Stratford and then on to Buckingham.

The intention was to canalise the river and climb the valley by a series of locks. The only artificial cut of any length was from Wolverton to Passenham, with the canal running in practically a straight line to Watling Street and then sweeping round in a gentle curve to Passenham church before rejoining the river. The course of the Ouse was then followed to Buckingham, with some of the worst bends being straightened by short cuts. By this means the engineering works were kept to a minimum and the two branches assured of a constant supply of water.

When the decision was taken for the main canal to cross the Ouse valley on an aqueduct the plans for the branches had to be

The Old Stratford branch left the main line immediately above Cosgrove lock, shown here in the spring of 1901. The swing bridge, which had just been erected, took the towpath to the branch across the lock chamber. *National Waterways Museum*

altered, and a new junction was selected at the north end of the embankment. As this point was nearly 40ft above river level at the aqueduct, the need for all but two of the locks on the Buckingham branch was eliminated. The new line now followed the contour up the valley until contact was made with the river for a short distance above Maids Moreton mill.

Considerable pressure was often put on the company by the towns to which branches were proposed for priority in the construction programme. Buckingham would probably have been well down the list had it not been for the Marquis of Buckingham, who lived near the town at Stowe. In May 1800 he agreed to lend the money needed to build the branch and asked for the work to be carried out as quickly as possible.

Once the labour force had finished cutting the main canal it went straight on to complete the Old Stratford branch. One and a quarter miles long, with two bridges, no locks and hence relatively cheap and easy to construct, it was completed in August 1800.

Work then started on the Buckingham branch. It was nine and a half miles long with two locks and took only eight months to complete. There was a formal opening ceremony on 1 May 1801, when a barge carrying the Marquis, with Praed and other committee members, led a procession of twelve barges carrying coal, slate and other goods into the basin at Buckingham, where they were greeted by a large crowd and by the firing of cannon[5].

The branches became of great importance to the town and the surrounding agricultural district, and were soon carrying 20,000 tons each year. One of the reasons for building the Buckingham branch was to export hay and straw for London's thriving horse-drawn traffic. In return, coal, stone and other goods were brought in to supply the area.

Paddington Branch.

As trade developed on the canal the company soon realised its connection with the Thames at Brentford was not ideal to attract much of the potential traffic to and from many parts of London.

To assist in building up trade through Brentford a wharf was opened on the Thames at Whitefriars, above Blackfriars Bridge. Land had been purchased in July 1794, and in May 1795 it was decided to build a covered wet dock and a two-storey warehouse. The works were completed and the wharf was opened early in 1796. Later in the year increasing trade led to an extra storey being added to the warehouse and to other improvements being made. By the early 1800s over £28,000 had been invested in the facilities. Despite the success of the wharf the benefits of extending the canal nearer to London were becoming increasingly apparent.

In May 1794 Jessop and Barnes surveyed a branch to London, and selected Paddington as a possible terminus. At the time it was ideally situated on the edge of the built-up part of the capital, and it also had the advantage that it could be linked to the nearest convenient point on the main canal at Southall by a level branch.

The plans were adopted in June and a Parliamentary Bill was promoted. Barnes completed detailed plans by the beginning of September and the Act was obtained in 1795 with relatively little trouble, as the scheme had overwhelming public advantages. The Paddington branch was soon to become by far the most important and successful of all the Grand Junction's extensions.

The upper section of the Buckingham branch, showing its connections with the Great Ouse.

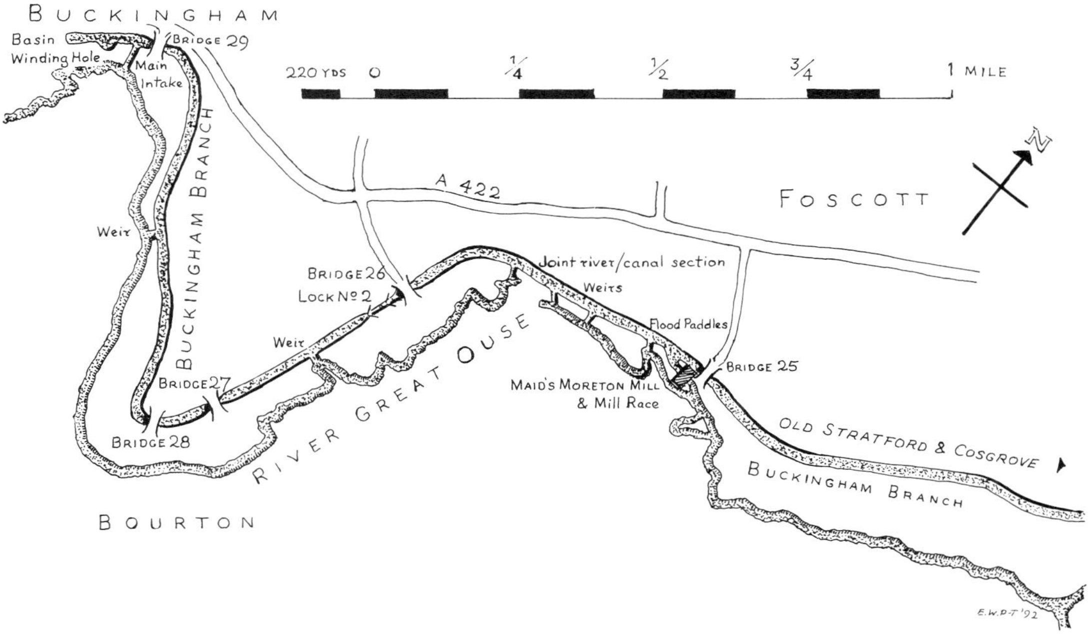

Although Jessop reported in June 1795 on the best way to build the branch nothing was done, as efforts and funds were then being concentrated on the main line. In June 1796 the company decided no money should be spent on branch canals unless funds were subscribed specifically for their construction[6].

The Paddington branch was seen as extremely important, and £5,000 was specifically reserved for work on a deep cutting and embankment over the river Brent near Alperton. Barnes and a surveyor named Gream then set out the line of the branch, and in October the company decided to raise £50,000 so that work could start without encroaching on the money for the all- important main line[7]. In November land purchases started, with construction beginning soon after. A special committee was set up in June 1797, under the chairmanship of Edward Dayrell, to supervise the work.

In November Barnes reported that over a mile between Horsendon Hill and the Brent valley was finished and full of water. Six floats were in use on this section ferrying earth for the embankment. Bricks had also been manufactured for the aqueduct over the river and for those bridges that were to be built next summer.

Whilst the building work was going well negotiations to acquire some of the land were not. By February 1797 there were major disagreements with the Bishop of London over land at Paddington, and this was the start of a protracted wrangle. In August, as matters were no further forward, a diversion from Kensal Green with an extension to Marylebone was being considered.

In September Parliamentary notices were given for three possible routes for the extension, and in November the proprietors authorised an approach to Parliament, in case agreement could not

The entrance to the Paddington branch at Bulls Bridge, Southall, in April 1958. Plans announced by British Waterways in 1955 to improve the entrance by easing the tight turn into the branch never materialised. *Railway & Canal Historical Society*

be reached with the Bishop and his tenants. This may have provided the spur, for in December an outline agreement was drawn up. Praed finalised this in April 1798.

Somewhat belatedly, Gream was instructed in May 1799 to purchase a 96ft wide strip of land wherever possible to allow for a towing path on both sides of the branch, as authorised in the Act. In October Barnes was told to provide the two paths, but by April 1800 both he and Jessop agreed that no useful purpose would be served by having a double path on the section not already completed. To save considerable extra cost it was ordered that a single path should be provided between Southall and the Brent aqueduct.

In June 1799 planning for the buildings at Paddington started, and before long the plans were so extensive that negotiations had to be re-opened with the Bishop to obtain more land. In March 1800 Barnes stopped all work southwards from the aqueduct until an agreement was reached, and instead used his labour force to complete the embankments and the other works on the branch.

Jessop reported in April that the branch from Southall to the Brent valley was finished. From there to Paddington work was progressing well. The only problems were the embankment near the river and the deep cutting immediately to the south. If over 200 men could be employed in these places the canal could be open to Paddington by Christmas.

Six months later much remained to be done and more bricks were having to be made at Alperton, but the end was in sight, and on 21 February 1801 Barnes was told to set out the line from the Harrow Road bridge to the terminus at Paddington. The committee met three days later to approve his proposals. Work started immediately afterwards, with the excavated soil being used to form the base of

The embankment across the valley of the Brent, the largest earthwork on the Paddington branch, was pierced by the North Circular Road in the early 1930s, necessitating this aqueduct over the new road. *National Waterways Museum*

the wharves around a terminus basin. At the same time the company advertised it was willing to let land at Paddington for wharves, warehousing and other purposes.

Work on the final stretch was soon finished, and the branch was opened on 10 July 1801 with full ceremony. At 9am the committee left Paddington in two pleasure boats bedecked with colours and streamers, and at noon they reached Bulls Bridge, where they were met by the City shallop and several other boats. The boats then returned in procession, reaching Paddington at 5.30pm, where they were greeted with the firing of cannon on Westbourne Green bridge and a volley of musketry[8].

Well before the branch was opened its potential for water supply purposes had been realised. The Duke of Portland, who was involved in supplying water to parts of Marylebone, made an approach in October 1794. Nothing came of the enquiry, but it prompted the company to investigate the position, as at the time many parts of London were without a piped supply. When an Act[9] was obtained in 1798 to confirm the agreement with the Bishop of London powers were included enabling the Grand Junction to become a statutory water undertaker.

In October 1799 a link with the Chelsea Waterworks Company was considered, but no agreement was reached. A second approach

A crowded scene in April 1909, when Paddington basin had been closed and drained for cleansing work. This necessitated moving all the boats out to what is now known as "Little Venice", where the Regent's Canal joins the Grand Junction. Paddington basin is approached through the bridge on the right. *National Waterways Museum*

in 1801 also failed, as did one from the New River Company in the same year and another from the West Middlesex Company in 1808.

Once the branch was opened the water supply business was developed in and around Paddington, and it soon became such an important sideline that it was transferred to a separate undertaking, the Grand Junction Water Works Company (GJWWC), by an Act[10] in 1811. This company remained in being until the Metropolitan Water Board was set up in 1902, when it was absorbed along with the other water supply concerns in London. Since 1820, however, the GJWWC had ceased to draw its supplies from the Paddington branch, apparently owing to complaints about its quality.

The scheme for the branch had a forerunner some years before the Grand Junction project materialised. In 1766 a canal was proposed from the river Colne at West Drayton to Marylebone, but soon afterwards the western end was amended to join the Colne at Uxbridge, from where links to other canals were visualised. For much of its course the proposed line was similar to that adopted by the Paddington branch. In 1773 an extension was proposed across the north of London to a large basin at Moorfields, on the site of present-day Finsbury Circus. A further extension to the river Lee at Waltham Abbey was also proposed[11].

Considerable thought was given to extending the Paddington branch still nearer to the City of London. Initially prompted by the dispute with the Bishop of London, the scheme was considered later on its own merits for its commercial advantages. In 1800 an extension to Tottenham Court Road was suggested, but the company decided the 30ft drop in level would give rise to too many problems.

In 1802 Thomas Homer, now a trader on the canal, investigated a canal from Paddington to the Thames at Limehouse[12], and a railway to the docks was also suggested. John Rennie carried out a survey, and whilst no action resulted at the time out of the idea grew the Regent's Canal project, authorised in 1812 and opened on 1 August 1820. It joined the Paddington branch at the Westbourne Green bridge.

At first the Regent's tried to obtain a supply of water from the Grand Junction. The talks were unsuccessful and an Act of 1816 authorised it to draw water from the Thames at Chelsea instead. In 1819 this supply was transferred to the GJWWC and in exchange the Regent's obtained water rights from the Grand Junction.

Prior to the opening of the branch Paddington had been a quiet rural community. It was transformed almost overnight into a thriving inland port. The company was quick to develop the land it owned and its property interests soon became a profitable sideline.

A large trade was soon passing over the Paddington wharves. In 1810 over 113,220 tons were received and 67,728 tons despatched by the branch, the most important inwards cargo being bricks needed for the rapid expansion of the capital at the time. At its western end the branch passed through good clay-bearing land and brickyards were established near Hayes and Southall, several of them served by private branch canals or docks. Other typical items were manufactured goods from the Black Country, stone from quarries in Leicestershire, and salt from Cheshire. Exports from Paddington included many of the waste products of the capital such as ashes, breeze and manure.

With the opening of the branch the commercial emphasis shifted away from Whitefriars, and soon afterwards Pickfords, the carriers, leased much of that installation. In December 1809 part of the wharf was sold, but the rest was retained as an investment property. When the Regent's Canal was opened some trade was diverted from Paddington to the City Road Basin, which was more centrally placed. Paddington, however, retained an important share, and the branch itself was just as busy as the diverted trade still passed along it on its way to or from the Regent's and the main line.

As soon as the branch was opened the company established a regular passenger service from Paddington to Uxbridge, initially with a single boat, but later with another two craft. They were named *Royal George, Marquis of Buckingham* and *William Praed.* Their crews were provided with smart blue watermen's jackets with yellow stand-up capes, yellow buttons and yellow badges on the sleeve lettered "GJC".

The boats made a useful profit. For the first 18 months there was a surplus of £2,709. The return then declined, and in May 1803 the company leased the boats to Henry Weeks for the rest of the year for £500 and 25% of the takings over £1,000. In January 1804 Homer leased the boats for three years at £750 per annum, and in 1807 the lease was renewed for another five years at £500 per annum. The regular services then petered out and thereafter the only passengers carried from Paddington were on excursion trips, which became quite popular at one time. Soon after the opening of the branch there was a private passenger and goods service from Paddington to Buckingham, but this did not last long.

Northampton Branch.

Although a branch had been authorised to Northampton in 1793 it was to be twenty-two years before it was completed. It had been intended that it would link up not only with the river Nene but also with the Leicestershire & Northamptonshire Union Canal (L&NUC) that was projected from Leicester via Market Harborough to Northampton.

In the deposited plans the branch was to have left the main canal some distance north of the present junction and run down a natural valley in a north-easterly direction to link up with the present line at the bottom of the Rothersthorpe flight of locks. In 1796 Barnes made a new survey, and in October he estimated £25,349 would be needed to construct the branch. The main items were 116ft of lockage at £12,760; 8,060yd of cutting and puddling £3,224; thirty-eight acres of land at £60 per acre £2,280; with the bridges, earthworks at the locks, culverts, and fencing making up the remainder.

This was higher than Jessop's earlier estimate, the difference caused by increased prices for materials and land and by making the locks 6ft longer. A railroad had also been suggested. Barnes did not comment on this apart from mentioning the obvious disadvantages of transhipment at either end. No action was taken and an estimate from Outram early in 1798 for a railroad from Blisworth to Northampton was also shelved.

Early in 1800 the commissioners in charge of the Nene navigation were becoming concerned at the loss of trade on their river, and they offered to raise the money to enable the canal branch to be built. As the company was concerned about its water supplies it would only agree to a railroad being built, provided Northampton raised the money. Jessop was told to carry out a survey.

His report was issued in April. He estimated the cost of a single-track railroad from Gayton to Northampton at £9,000. As the line could be out of action for weeks at a time, owing to floods in the low-lying meadows crossed near the town, he recommended a part canal, part railroad, scheme instead.

From Gayton he suggested a level canal about five-eighths of a mile long; then a double-track railroad falling on a 1 in 72 gradient; and finally a canal to the town by the Parliamentary line for two and three quarter miles, with three locks each having an 8ft fall. These locks were to be 100ft long and 10ft wide to accommodate two Nene barges.

His estimate was £12,975. Of this £1,040 was for the upper canal, £7,123 for the lower canal, £3,120 for the railway and cranes, and the balance for contingencies. A single-track railroad would reduce the cost to £11,855. Jessop envisaged one horse would be able to work the railway, and that it could deal with 20,000 tons each year.

The Nene commissioners approved Jessop's plan in April only reluctantly, as they would have preferred a canal all the way. In May

The course of the Northampton railroad at Gayton.

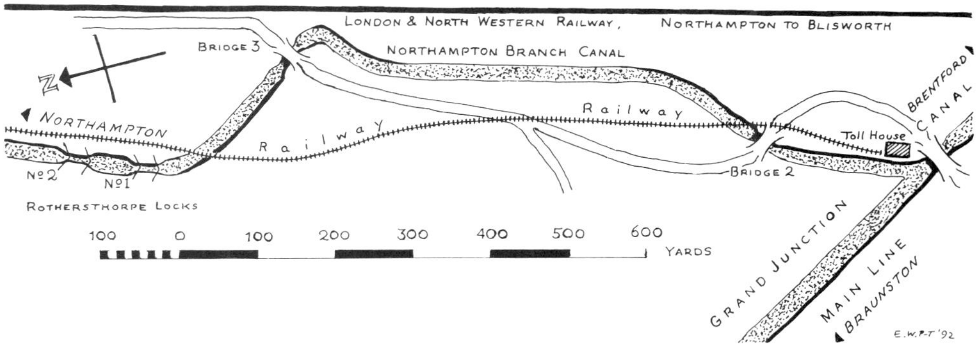

The distinctive lock cottage alongside Rothersthorpe top lock on the Northampton branch.
Robert Wilson

they opened a subscription list to raise £12,000[13], but the money was not forthcoming. As a result the company took no action.

In October 1802 a strong resolution in favour of the canal being cut was passed at a public meeting in the town, chaired by Lord Northampton[14]. At the time the company was applying for an Act to raise more money, but Northampton was opposing this unless the company agreed to build the branch.

In November, when the public meeting was resumed, the company offered to build the upper canal from Gayton and then a railway down to Northampton by June 1804[15]. The meeting rejected this plan, demanding instead the canal as specified in the Act. This demand was in turn rejected by the company, which was still concerned about the water supply for a branch canal.

The Nene commissioners had suggested that reservoirs could overcome the water shortage, but these were not considered satisfactory. Further discussions followed, and in February 1803 agreement was reached for the combined canal and railroad to be built by 25 March 1805[16]. Sufficient funds were to be earmarked from the money being raised by the company to complete its other works. This provision was inserted in the Act.

Barnes and John Barker, another surveyor, then drew up detailed plans for the work. At first the idea of a railroad linking two sections of canal was revived, but in August 1803 a level canal from Gayton and a single-track railroad down to Northampton was specified, although sufficient land was to be taken for a double line. In November 1804, in an attempt to keep the cost down, Barnes was told to make the line a railroad throughout.

Despite the time limit imposed by the 1803 Act little was done until the end of 1804. The delay was deliberate, as after the opening of Blisworth tunnel in March 1805 the materials from the double-track railroad over the top of the tunnel became available. They were then used for the Northampton line, supplemented by other rails the company had been using at Wolverton.

Once started, work on the new railroad proceeded rapidly, and it opened on 7 October 1805[17] at a cost of about £12,000. The line followed the course of the present-day canal except at Gayton, where a more direct course was taken from the top lock to the main canal. At the junction the line terminated alongside a small basin built to handle the Northampton traffic.

The railroad did not satisfy Northampton, and when the company obtained another Act in 1805 authorising increased tolls a clause was inserted exempting traffic to Northampton until the branch canal was completed.

The townsfolk continued to complain, probably because the railroad was not ideal for all types of goods. It was satisfactory for coal but inconvenient for such things as grain and timber. In May 1807 two road carriers established a service from Blisworth wharf to Northampton. In 1809 Northampton Corporation claimed the railroad was inadequate as it was less convenient and more expensive than a canal, and articles were subject to breakage and pilferage.

At the end of 1809 the promoters of the Grand Union Canal (GUC), which was to link the unfinished Leicestershire & Northamptonshire Union Canal (L&NUC) with the Grand Junction at Norton, were becoming active. In December Northampton decided to oppose the new Bill as it would finally rule out any possibility of the L&NUC ever completing its extension to the Nene. At the same time the town renewed pressure on the Grand Junction, one of the main sponsors behind the GUC project, for the early completion of the branch canal from Gayton.

The Grand Junction capitulated in January 1810 and agreed to complete the branch in three years[18]. Northampton then withdrew its opposition to the GUC Bill, but as the branch was planned with narrow locks, to save money and water, a clause was inserted that if ever the locks on the GUC were widened those on the Northampton branch must also be widened. The Nene commissioners agreed to help with financing the work against a Grand Junction guarantee of 5% interest on the sum lent.

Nothing was done until June 1812, when it was estimated that £35,000 would be needed to complete the branch. Soon afterwards stocks of bricks began to be collected, the exact sites of the locks were worked out, and plans made to cause as little interference to the railroad as possible. In August 1813 Benjamin Bevan, who was in charge of the works, reported that the foundations of the bottom lock were nearly complete and the canal across the meadows in Northampton to the next lock was cut and puddled. Bricks were ready on site for seven locks, with some provided for another two. A further million bricks were in store at Gayton and more were being made. Half of the oak timber had been provided and 20,000ft of sawing done. Two lime kilns had been built and were in production.

A lift bridge at lock No 10, Rothersthorpe, on the Northampton branch.

By June 1814 completion was forecast for October. Over three and a half miles of cutting had been completed, seven locks were finished and the bricks for all the rest were on site. All the lockgates had been framed, and the ironwork finished on several of them. In November, however, much still remained to do. Another half mile of cutting and another seven locks had been completed. All the lockgates were finished, but only a third had been hung and fitted. Seven bridges were finished, but six still remained. The revised opening date was the beginning of March 1815.

Even this was optimistic, for it was not until 1 May 1815 that the branch, just under five miles in length and with seventeen locks, was finally opened. A great crowd gathered to see a procession of twenty boats arrive carrying coals and merchandise from various parts of the country[19]. The branch was an immediate success. Grain to mills on the Nene and leather goods and strawboard for the shoe trade in Northampton were important traffics.

The cost of the branch was £35,382. Little is known about the fate of the railroad, but in December 1816 some wooden blocks taken from the line were sold. In 1824 the company offered other materials to the Thames Tunnel at £4.20 per ton. The rails had been removed to Gayton and in 1824 they were being offered at £6.50 per ton.

A view from the final lock on the Northampton branch on 15 May 1957 showing the junction between the canal and the river Nene in Northampton. The buildings behind have now been demolished and the site has been covered by a modern lager brewery. *Alan Faulkner*

Aylesbury Branch.

Aylesbury found itself in a very similar position to Northampton as it was many years before its branch, authorised in 1794, was built.

No action was possible until the main canal was open to Marsworth in 1799. In August 1800 people in Aylesbury offered to raise enough money to build a railroad to the town. The company agreed and ordered the cast-iron rails, but in November a change of plan saw the rails diverted to Wolverton, where they were later used in the construction of the embankment.

The reason for the delay was a suggestion to extend the canal beyond Aylesbury to Abingdon, and a group of subscribers in the Wilts & Berks Canal (W&BC) had commissioned a survey. At first the Grand Junction was only mildly interested as it would have to provide the necessary water. The main canal was not well supplied at the time, which is why the company was keen to adopt a railroad.

In April 1802 Barnes re-surveyed the line of a railroad to Aylesbury and in May another surveyor, John Holland, started work between Aylesbury and Broughton mill. Little can have been done, for early in 1803, following a report by Jessop, it was decided it would be better to have a canal[20]. Barker was now asked to survey a canal line and find some means to save water by reducing the lockage. Barnes and Holland assisted in this work. Despite this nothing else was done.

A wooden footbridge on the Aylesbury branch at Wilstone.

People in Aylesbury were now becoming concerned at the lack of progress. Then in February 1805 the company agreed to build a canal, or a railroad, to the town but only if extra funds could be raised, as the original £20,356 offered was no longer sufficient.

In November a group of the original subscribers threatened legal action, but the company would not be moved. In the autumn of 1806 a Bill was introduced into Parliament to force the company to build the canal[21] and to use the promised money towards the work, but it failed.

In 1808 the Marquis of Buckingham proposed a canal to run from Seabrook, not far from Marsworth, to Thame. The company agreed to support the project with a £15,000 loan spread over three years but imposed various conditions, mainly over the all-important water supplies. On learning of these the Marquis decided not to proceed.

A link with the W&BC came up again in October 1809. The Aylesbury to Abingdon scheme, now known as the Western Junction Canal (WJC), was strongly supported by Aylesbury. Its opportunity came, like Northampton's, with the Grand Union Bill. The Marquis and several people in Aylesbury petitioned against this and agreed to withdraw their opposition only if the company undertook to build the Aylesbury branch.

An agreement was finalised in March 1810. If both the GUC and WJC Bills passed through Parliament, the company would allow the WJC to make the branch from Marsworth to Aylesbury as part of its line to Abingdon. If the WJC Bill failed but the GUC Bill passed, the company agreed to build the branch within an agreed time. If neither Bill passed the company agreed to construct a railroad to Aylesbury, unless a subsequent agreement was made with the Marquis for a canal to Thame.

At the same time the company reached agreement with the W&BC over the WJC plans. This line, as surveyed by Barker, Provis and William Whitworth, was to run from Marsworth to Aylesbury and then through Stone, Long Crendon and Cuddesdon to join the W&BC near Abingdon. The estimate was £200,000, of which £70,000 each was to be found by the Grand Junction and W&BC shareholders and the remaining £60,000 from the general public. Water was still a crucial factor and the Grand Junction insisted no lock should have a fall greater than 6ft.

There was considerable opposition from the Kennet & Avon Canal (K&AC), which was promoting a rival canal from near Maidenhead to Cowley. It originated in August 1795 in a scheme for a canal from Cowley to Boulters Lock to bypass a treacherous stretch of the river Thames.

Little more was heard of the plan until the end of May 1801, when Holland was asked to survey a canal from Cowley to the Thames at Marlow, with a branch to Reading. Holland worked in collaboration with Barnes, whilst Peter Potter was appointed to assist with the surveying and measuring.

Holland made three proposals: a canal from Cowley to the Thames near Marlow; another for a canal from Cowley passing

A narrow boat discharges grain at Walton mill, just above the bottom lock in Aylesbury. In 1927 canal craft brought to the town 5,664 tons of grain, 1,181 tons of timber, 3,673 tons of coal, 496 tons of ballast and 275 tons of sand. *Railway & Canal Historical Society*

Narrow boats unloading at a factory on the last stretch of the canal in Aylesbury, near the terminal basin.
National Waterways Museum

through Maidenhead to the Thames at Sonning; and a third from Cowley to the river Kennet at Reading. In November the third scheme was chosen and a Bill submitted to Parliament.

Opposition came from landowners whose property would be affected, and from the Thames Navigation Commissioners, who were largely landowners themselves. Several well-supported opposition meetings were held, and faced with this the Bill was withdrawn. It was hoped the landowners would change their minds, but there had been a marked lack of support from the K&AC company, which was fully occupied in trying to complete its own canal.

In 1810, and with its own line now complete, the K&AC produced its own plans. Again there were three canals: from Cowley to the Thames at Boulters Lock and then from Cookham to the Kennet at Reading; from the Thames at Isleworth to the Kennet at Reading; and from the Basingstoke Canal, near Basingstoke, to its canal at Newbury.

The Grand Junction was asked for its support but wanted to know which scheme was to be adopted and what the landowners' reactions were. No further action was taken, but the Cowley to Boulters Lock scheme continued to be discussed and became known as the Western Union Canal (WUC).

The defeat of the WJC Bill meant Aylesbury was no nearer getting its canal. In March 1811 the Marquis of Buckingham and others interested met the company, when it was agreed work on the branch should start and be financed by calls on the WJC subscribers. This money was to be regarded as a loan, and if Parliament authorised the WJC Bill it could be converted into shares and the Aylesbury branch was to be handed over to the new company. A £20,000 limit was set on the money, with the agreement expiring in August 1812, when the company was liable to construct the branch as agreed in

March 1810. In August Thomas Tindall, a solicitor in Aylesbury who had been very active in trying to get the branch built, paid over the first instalment of the loan and on 13 August Henry Provis was ordered to start the work.

Bridge No 2 on the Aylesbury branch at Long Marston.

Two years later he reported that nearly a mile of canal from Marsworth was navigable through seven locks. A further three and a half miles through four locks was complete at the Aylesbury end. Most of the remaining cutting had been started and work had begun on two of the remaining five locks. Ten of the eighteen bridges had been finished and two more were nearly complete. Up to the end of July 1813 nearly £24,000 had been spent, and a further £18,000 would be needed. This excluded the cost of the land.

The actual completion date of the branch is uncertain. One source suggests March 1814[22], which is feasible assuming the same rate of progress from the time of Provis's report. A report in November 1814, however, stated the work was not finished. Soon afterwards Telford was asked to inspect the branch to help answer criticisms about the way it had been built. He reported in May 1815 that it was "in a very perfect state"[23]. He seemed very impressed with Provis and his work.

One factor bearing on the opening date was the drawings of money made by Provis. Throughout 1813, when the bulk of the work was being done, he drew over £1,500 each month to pay for men and materials. This dropped steadily in 1814 and finished altogether in June 1815. Another factor was the critical water supply situation. New reservoirs were built on the Tring summit at Tringford and Startopsend, specifically for the Aylesbury branch. Tringford was started early in 1814 and Startopsend in 1815. Without them the supply to the main canal would have been very strained. It seems likely that work on the branch was deliberately slowed down in 1814 so that work on the new reservoirs could catch up, with the branch being opened early in 1815 when the water position allowed.

Meanwhile the WJC proposal was not dormant. In September 1813 notices appeared for an Aylesbury & Abingdon Canal to run from Aylesbury to the Thames at Culham, with a branch to Wotton Underwood to improve its water supplies. There were to be reservoirs at Stoke Mandeville and Ellesborough. A second branch was suggested to Chalgrove, again to act as a water feeder.

At a public meeting in September at Thame, chaired by Lord Nugent, an estimate of £191,800 was given[24], excluding the Chalgrove branch. A committee was appointed but there was no mention of taking over the Aylesbury branch as had been intended in March 1811. Like the WJC it came to nothing.

The WUC project was revived in 1815 when John Rennie made a survey. His plan was approved by the Grand Junction in August 1816, providing the level of the new canal at Cowley was higher than its own level. There were talks with the promoters in September, but no action resulted.

Three years later the K&AC company revived the plan. The Grand Junction approved the scheme again, but with the same safeguards about water levels at Cowley. It also insisted on a surcharge on all iron entering from the new canal, so as to protect its Staffordshire iron trade. A prospectus for the WUC was issued which showed that the twelve-mile canal, with a branch to Windsor, would cost £130,000.

There was considerable opposition when the Bill was presented to Parliament. Foremost were the Thames Commissioners and the W&BC company, which had revived its WJC plans early in 1818. Telford had surveyed this, and the Grand Junction given its approval, but with compensation tolls similar to the WUC.

In May 1820 the Grand Junction decided both schemes would be to its advantage, but if the W&BC and the K&AC companies opposed each other they would prevent either scheme being carried out, as there was insufficient trade to justify both canals being built.

Despite the opposition the WUC Bill passed its second reading in the House of Commons in May 1820. The Grand Junction then decided to increase its surcharge on iron. The Bill now ran into difficulties in the committee stage largely on account of this surcharge, to which Lord Shaftesbury took particular exception. The company would not relent and partly for this reason, and partly due to continuing opposition, the Bill was withdrawn in July. Effectively this was the end of the WJC and WUC projects, although both were reconsidered in various forms in the next few years.

A motor boat loading wheat at Brantom's dock in Leighton Buzzard for delivery to London in 1928. This dock was one of the earliest on the canal, being built by John Grant in 1801. It was filled in during 1974, but the bridge carrying the main canal towpath across the entrance still stands. *Andy Collier*

The canal entrance to the fortified part of Weedon Military Dock on 26 September 1961. This section of the branch has been filled in and built over, but the gatehouse, a listed building, survives. *Alan Faulkner*

The Military Docks.

In the early 1800s there was a real threat that the French Emperor Napoleon would mount an invasion against this country. The Government took a series of defensive measures, including the establishment of a large fortified ordnance depot at Weedon in Northamptonshire close to the canal. The plans included a 600yd branch canal, and work on this started in June 1805 and was completed a year later[25].

In February 1806 an extension of the branch to almost double its original length was authorised to serve ammunition stores and magazines about to be built. The extension and improvements to the first section were completed by the beginning of 1810[26].

A feature of the branch was a series of portcullises under which boats had to pass to enter, and leave, the fortified part of the depot and the magazine compound. These were raised and lowered by hand operated machinery in the buildings above. The branch was used mainly for the supply of stores, arms and ammunition and considerable quantities were shipped in, and out, by canal boat.

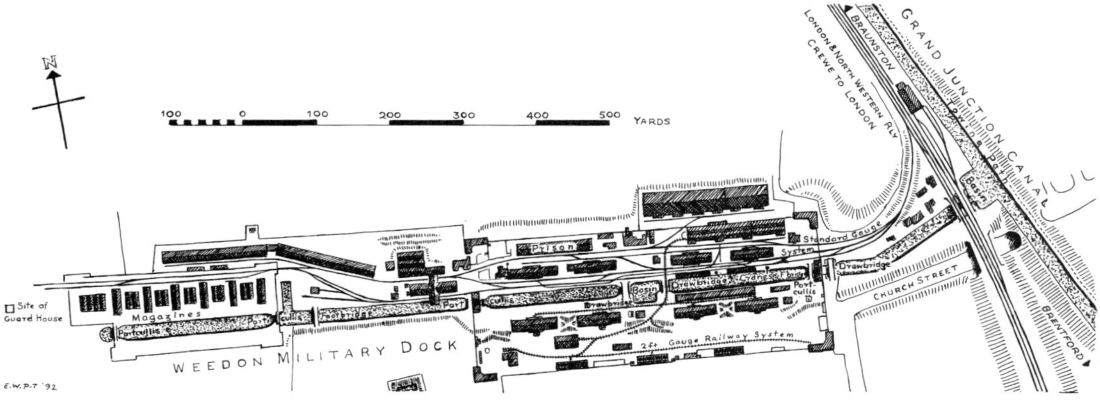

Weedon Military Dock, showing the full extent of both the canal and railway systems.

Several bridges crossed Weedon Military Dock within the fortified part of the depot. This former drawbridge was operated by filling a counterbalancing tank with water, the weight of water causing the bridge to rise; when the water was released the bridge closed. *Alan Faulkner*

Another depot was established at North Hyde, near Southall. Here a 1,200yd branch canal was intended to form a defensive loop around the munitions storage area, with just one road access bridge. Work started in October 1813[27] but ran into problems when large beds of gravel were encountered. A stop lock and special culvert had to be built and, as a result, the loop was not completed, with the branch stopping a few yards short of the second intended junction. The work was finished probably by the end of 1815.

The depot itself was completed by the end of 1817, and considerable amounts of gunpowder were brought in by barge from Waltham Abbey and Woolwich for storing on the site.

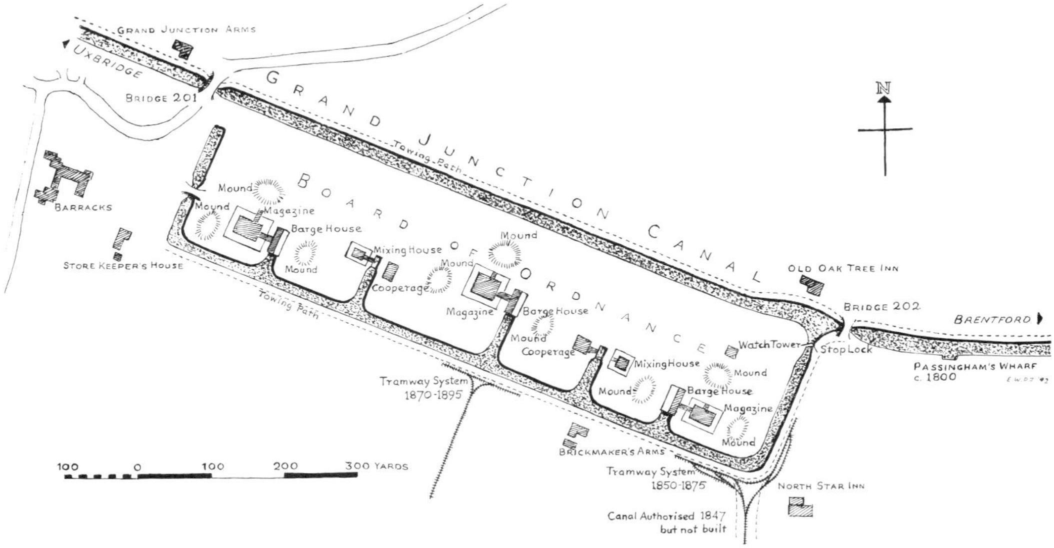

North Hyde Military Dock.

Other branches and docks.

Several branches were built at Rickmansworth. The first left immediately below Batchworth lock and ran for nearly 540yd up to a new wharf in Rickmansworth. It was a canalisation of the river Chess through one lock and was built for Salters brewery, which was sited close to the new wharf. Opened in 1804[28], it was often known as Salters Cut, and its main traffic was beer in barrels to Uxbridge, with the empty barrels in return. Coal and general goods were also brought to the town wharf.

Another branch to nearby Batchworth mill was about 150yd long. It was built by John Dickinson, the mill owner, and opened in August 1818[29]. Another, from just below Batchworth bridge, running over 300yd northwards into the Bury Grounds, was built for John Taylor, a baker, to enable flour to be carried up to his bakehouse. Halfway up its course the branch is crossed by a section of the river Chess.

Further south a branch was built at West Hyde to serve Troy mill and some chalk pits. Over 1,000yd long with no locks, it was built probably in 1800. It was certainly open in 1822.

At Leighton Buzzard John Grant applied to make a dock nearly 50yd long into his premises in December 1800. This was one of the first of many such requests the company was to receive and it involved breaching the towpath. Permission was usually granted, as

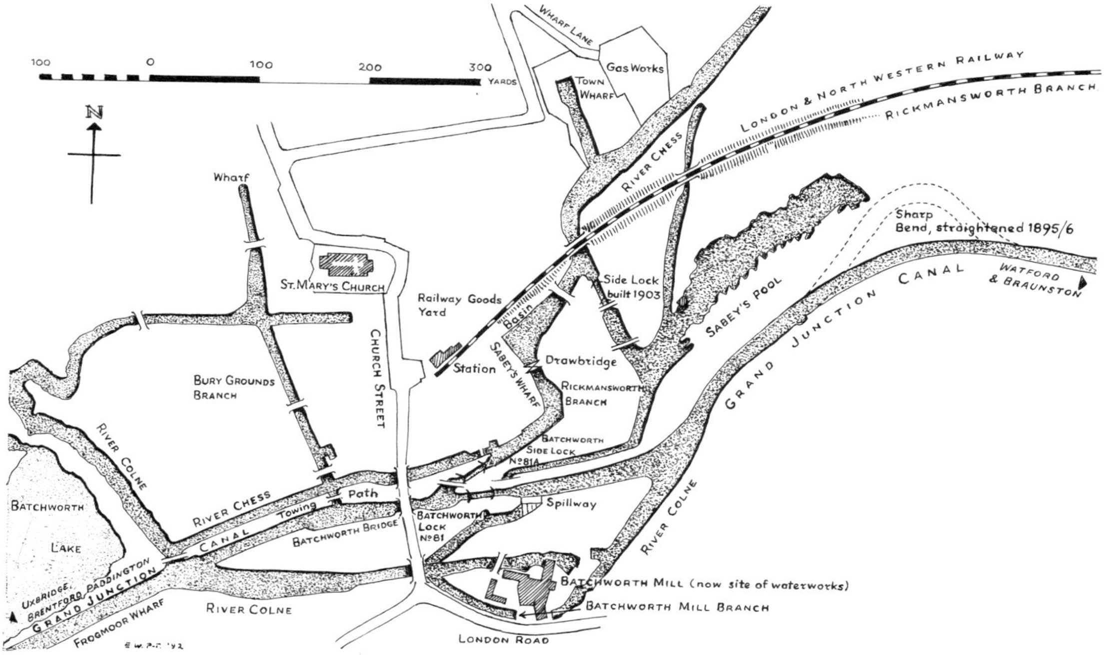

The Rickmansworth branch canals.

Stoke Bruerne mill, with the entrance bridge to its dock in the background. The mill, dock and four cottages were purchased by the Grand Junction for £350 in April 1913. The main mill building now houses the popular Waterways Museum, whilst the building to the left was later demolished; the site is now occupied by the museum's shop and offices. *National Waterways Museum*

in Grant's case, providing the company's engineer was satisfied the dock was watertight and an adequate bridge was erected if the towing path was being cut through. A nominal annual rent was usually charged.

At Braunston a 230yd branch was opened in August 1816 to serve a warehouse built by the company alongside the Daventry road. At Stoke Bruerne James Ebbern built a 50yd dock, opened in May 1842, to serve his mill by the top lock. In March 1846 George Savage was allowed to make another dock just over 100yd long into nearby land owned by the Duke of Grafton, for whom he was agent. This served the Stoke Brick & Tile Works.

A whole series of short branches or docks was opened between Uxbridge and Norwood. One of the earliest was the 330yd branch to serve the Treaty wharf at Uxbridge, opened in 1798. Just south of Cowley lock the 350yd long Cowley Hall Dock was opened in 1811, whilst at Cowley Peachey a short dock was opened in the early 1800s to which the Paddington packet boats ran. It is still known as Packet Boat Dock and The Paddington Packet Boat Inn stands on the Uxbridge road nearby.

In March 1818 work started on building a dock at Yiewsley to open up supplies of clay for brickmaking. This, known as Otter Dock, was completed in 1820 and after several extensions was eventually over a mile long. Another dock into a brickfield was opened between 1811 and 1822 at West Drayton. Later it became known as Pocock's Dock, and after several extensions it was over 1,100yd long.

They were typical of a series of other docks built into brickfields, particularly in the Dawley and Hayes areas, to help supply the growing London market.

Unsuccessful branches.

The GJC maintenance workshops at Bulbourne, which served the whole of the canal.

Many branches failed to materialise. The Chesham branch was projected to follow the course of the river Chess from Rickmansworth. Chesham was a small market town at the time and the expense of building a nine-mile branch rising 150ft through over twenty locks can hardly have been justified.

The Daventry branch was to have been one and three-eighth miles long, rising 50ft to Daventry through nine locks. It was mainly intended to supply water to the Braunston summit, for the town was hardly big enough to justify a branch otherwise. Later part of the line was taken for the large Daventry reservoir and this removed the need for the branch.

The four-and-a-quarter-mile Dunstable branch was planned to run in a straight line from Puddle Hill, just north of Dunstable on Watling Street, to Slapton and roughly parallel with the river Ousel. Opposition from landowners put a stop to the scheme both in 1794 and when it was revived later. As a result Dunstable had to be content with being supplied from a wharf at Bulbourne.

The one-mile branch to Hemel Hempstead was intended to follow the river Gade. The town petitioned unsuccessfully to have it built. Although Hemel had a busy market the company probably felt it was too small to warrant the expense of the branch. The idea was revived in the 1840s but nothing was done.

The Watford branch was intended to be a two-and-three-quarter-mile canalisation of the river Colne. The town was sufficiently important to justify a branch, but when the course of the main canal was diverted to pass much nearer the need became less obvious. The Earls of Clarendon and Essex tried in 1796 to get the branch built but the company would only agree if Watford raised the necessary finance. The Earls renewed their request three years later, but a vital factor was the restriction on carrying coals southwards on the canal (see Chapter 5). The company suggested a railroad instead, but this did not suit their lordships and the matter was dropped.

The lack of action on the Watford branch also affected that to St Albans, which was to have followed the course of the rivers Colne and Ver for about eight miles. Opposition from landowners forced the company to spend nearly £3,000 in surveying and Parliamentary fees. In the Act an additional 1p toll was authorised for all goods on the branch until these expenses were recouped.

The project had originated in the autumn of 1792 when Lord Verulam and the Mayor of St Albans had contacted Praed about their fears the town would be left with no outlet to a waterway[30]. A plan by Barnes was approved at a public meeting in the town on 26 August and a deputation sent to seek its acceptance by the company. Meanwhile a subscription was opened to defray expenses.

It may have been the extra toll, which would have deterred traffic, or the continuing opposition of several millers that made the company reluctant to proceed, despite continuing encouragement from Lord Verulam. In July 1808 Praed confirmed the company had no intention of going ahead.

In 1816 a railroad was proposed instead of the canal. It was to run from the nearest point on the main canal at Belswaines, just south of Hemel Hempstead, to the town. The company supported this project, which was to be carried out by a separate company, and offered the iron rails then available from the Northampton railroad. In February 1817 it agreed to subscribe £300 towards the cost of the additional rails needed.

Despite this the financial support was not forthcoming from St Albans and the project died. A proposal for a canal from Ivinghoe to Luton in 1813 also came to nothing.

There were several plans to link the Great Ouse with the Grand Junction. Early suggestions were for canals from the river Ivel, a tributary of the Ouse, at Shefford to Fenny Stratford and from Bedford to Fenny Stratford by two possible routes.

In November 1811 a meeting in Bedford considered a canal to the Grand Junction and Francis Giles carried out a survey, acting under orders from John Rennie. Two possible lines were again suggested: from Bedford to Woughton, north of Fenny Stratford; and from Bedford to Soulbury, south of Fenny. The Woughton line was chosen and a prospectus for the Bedford Canal issued in July 1812.

The line was to be fifteen miles long with twenty-five locks all falling to the Ouse. Nineteen were in a flight near Brogborough, where a 1,220yd tunnel was needed. The estimate was £180,807. Traffic was calculated between 60,000 and 70,000 tons each year, with corn from East Anglia passing to Birmingham and Manchester and coal and manufactured goods being received in exchange[31].

There was considerable opposition from landowners, and although the Grand Junction offered to subscribe £12,000 it later reduced this to £3,000, which proved a serious setback. This action was prompted by a report received from Bevan about the poor state of the river Ouse. Although £25,000 was raised locally this was not nearly enough and the plan collapsed in 1815.

A suggestion had also been made for a link from the Bedford Canal near Marston Moretaine to a proposed London & Cambridge Junction Canal to open up a direct line to Cambridgeshire and Hertfordshire.

In 1814 there was a plan to canalise the Ouse from Newport Pagnell downstream to Sharnbrook, from where a canal was proposed to St Neots to eliminate a very sinuous section of the river[32]. Heavy engineering works were involved. The Grand Junction would be reached through a short link to the Newport Pagnell Canal, then about to be built. There was much opposition from Bedford, which would have been bypassed, and the scheme failed.

In 1824 Rennie and Giles revived the plan for a canal from Shefford to Fenny Stratford, whilst in 1838 and again in 1844 attempts were made to revive the Bedford Canal on a slightly different line. Giles and Bevan did surveys, but all without success.

A final proposal came in 1892 for a Bedford & Grand Junction Canal. This proposed a seventeen-mile canal from the Ouse at Kempston, near Bedford, to Soulbury through twenty-five locks[33]. A company was formed which was to have a capital of £360,000 and be empowered to buy the Ouse navigation from Bedford downstream. The plan to take over the river caused much local authority opposition and the Parliamentary Bill was withdrawn in July 1892.

NOTES TO CHAPTER FOUR

1 34 Geo III, c 24—28 March 1794.
2 35 Geo III, c 43—28 April 1795.
3 35 Geo III, c 85—2 June 1795.
4 General Committee minutes, 15 March 1797.
5 *Northampton Mercury,* 9 May 1801.
6 General Assembly minutes, 7 June 1796.
7 Ibid, 1 November 1796.
8 *Northampton Mercury,* 18 July 1801.
9 38 Geo III, c 33—26 May 1798.
10 51 Geo III, c 169—15 June 1811.
11 Spencer, Herbert. *London's Canal,* 1961.
12 Ibid.
13 *Northampton Mercury,* 3 May 1800.
14 Ibid, 23 October 1802.
15 Ibid, 27 November 1802.
16 Ibid, 18 February 1803.
17 Ibid, 5 October 1805.
18 Ibid, 13 January 1810.
19 Ibid, 6 May 1815.
20 General Committee minutes, 17 March 1803.
21 *Northampton Mercury,* 13 September 1806.
22 Gibbs, Robert. *History of Aylesbury,* 1883.
23 General Committee minutes, 11 May 1815.
24 *Jackson's Oxford Journal,* 2 October 1813.
25 Compton, Hugh & Faulkner, Alan. Weedon Military Dock, *Railway & Canal Historical Society Journal,* November 1991.
26 Ibid.
27 Compton, Hugh & Faulkner, Alan. North Hyde Military Dock, *Railway & Canal Historical Society Journal,* July 1992.
28 Rickmansworth Vestry minutes.
29 Evans, Joan. *The Endless Web,* 1955.
30 Lansberry, H.C.F. St Albans Canal, *Hertfordshire Past & Present,* 1967.
31 Bedford County Record Office, GA 736/2 & 3.
32 BCRO, GA 736/1.
33 Deposited Plan. Bedford & Grand Junction Canal, 1892.

The Early Days 5

AFTER THE engineering problems, finance was the most pressing matter confronting the company in its early days. The original Act[1] authorised £600,000 being raised. Calls of £10 on each £100 share were made at regular intervals; for instance in September and December 1793, and in March and July 1794. By July 1795 it was obvious more money was going to be needed.

At the half-yearly meeting in November the shareholders agreed to an application to Parliament for power to increase the capital. The Act[2] was obtained quickly; it authorised another £225,000, raised by issuing 4,500 half shares of £50 each to the existing shareholders. Two half shares entitled the holder to one vote. The Act also authorised an additional toll of 0.1p (one farthing) per ton mile on lime, limestone, ironstone, bricks, tiles, slate, coal and manure, with an additional 0.2p (one halfpenny) per ton mile on all other goods.

The increased capital was still not enough, particularly as the company was going to have to spend heavily to develop its water supply business in Paddington. A further £150,000 was authorised by an Act[3] in 1798. This was raised on a loan paying interest at 5% and convertible into shares at par in the early 1800s. It became known as the First Optional Loan and attracted £165,000.

From 24 June 1798 the payment of interest on the share capital was stopped as an economy measure. Until then 5% was being paid and this had cost the company £92,404. Instead, and as the tolls for the three years to 29 September 1798 had yielded over £15,000, income was now to be applied towards paying a dividend. Another economy measure had been to discontinue Jessop's appointment as chief engineer in September 1797; he was retained as a consultant instead, and continued just as busy in the company's affairs.

Capital was still short, and in the autumn of 1800 further powers were sought. In 1801 an Act[4] authorised another £150,000, bringing the capital up to £1,125,000. It permitted the £100 and £50 shares being subdivided into shares of £12.50 if required, but this was not carried out.

This Act also amended the agreement with the Corporation of London. From the passing of the original Act to the end of 1798 the company had to make good a £1,562 deficiency in the compensation toll. In future the company was to pay £600 each year, but the right to the toll on all goods passing between the Thames and the canal was transferred to it.

It was becoming difficult to raise more capital from the long-suffering shareholders and further loans had to be arranged instead. A second optional loan raised £100,000, a third £28,000

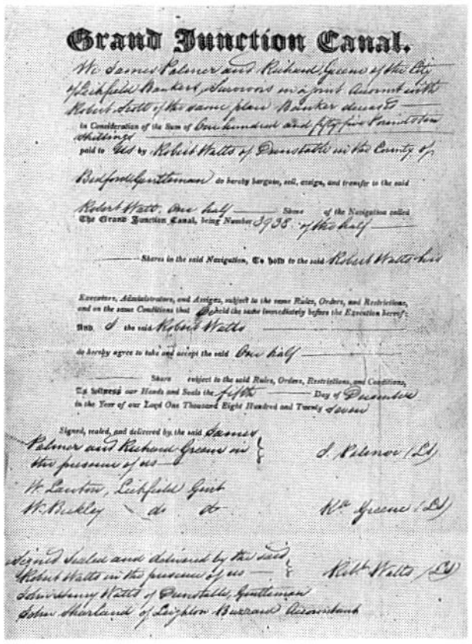

Grand Junction Canal.

We James Palmer and Richard Greene of the City of Lichfield Bankers, Survivors in a joint Account with Robert Scott of the same place Banker deceased

In Consideration of the Sum of One hundred and fifty five Pounds ten Shillings paid to Us by Robert Watts of Dunstable in the County of Bedford Gentleman do hereby bargain, sell, assign, and transfer to the said Robert Watts One half Share of the Navigation called The Grand Junction Canal, being Number 3938 of the half Shares in the said Navigation, To hold to the said Robert Watts his Executors, Administrators, and Assigns, subject to the same Rules, Orders, and Restrictions, and on the same Conditions that We held the same immediately before the Execution hereof: And I the said Robert Watts do hereby agree to take and accept the said One half Share subject to the said Rules, Orders, Restrictions, and Conditions. As Witness our Hands and Seals the fifth Day of December in the Year of our Lord One Thousand Eight Hundred and Twenty seven

Signed, sealed, and delivered by the said James Palmer and Richard Greene in the presence of us — J. Palmer (L.S.)

W. Lawton, Lichfield Gent

W. Bukley do do — Rd. Greene (L.S.)

Signed Sealed and delivered by the said Robert Watts in the presence of us — Robt. Watts (L.S.)

John Henry Watts of Dunstable, Gentleman

John Sharland of Leighton Buzzard Accountant

A Grand Junction share transfer form, dated 5 December 1827, in respect of one half-share of £50. *Bedfordshire County Record Office*

and a fourth £80,920, all at 5% and with rights to convert into ordinary shares in 1808. In addition £39,990 had been raised on mortgage.

By the beginning of 1802 it was obvious that still larger sums were needed, and the company was once more fast approaching the limit Parliament had allowed it to raise, there was growing criticism and concern at the canal's insatiable appetite for funds, and an in-depth review was conducted in June.

So far £1,104,610 had been promised in shares and loans but only £1,039,909 received, which included £32,435 from the sale of surplus land. Up to the end of May expenditure amounted to £1,067,023. The deficit had been financed partly with a £12,000 loan from the treasurers and partly from the company's growing revenue. There were outstanding capital debts of £74,615 which the company hoped to pay by selling more land surplus to its needs and from the £60,000 still owing on the shares and loans.

Meanwhile Barnes produced a long report[5] showing why the works had cost so much more than the estimate. Basically there had been an all-round steep increase in prices since the estimate was made in 1792. This inflationary rise was caused mainly by the Napoleonic wars.

A labourer's wage had gone up from 10p to 17½p per day; the price of oak from 8½p to 11p per foot; elm from 4p to 6p per foot; and fir from £3.50 to £6.75 per load. In 1793 a waggon and five horses cost 90p per day to hire; this had now doubled to £1.75. The rise in the price of labour was estimated at ½p per day for 800 men for eight years, totalling £99,840; the increased cost of waggons, horse teams, iron, timber and an additional duty on bricks almost £44,000; whilst the price of land had risen on average by £25 per acre, involving £22,500.

Then the company's Act had imposed several conditions not allowed for in the estimate. Before spoil banks were started the turf and 9in of top soil had to be removed, and replaced afterwards. The company had been compelled to purchase certain odd bits of land no longer wanted by the owners where they had been split up by the canal. These two alone had cost £34,000.

The whole scope of the undertaking had also grown. The original estimate had provided for short branches to Daventry and Watford, which had been replaced by much longer branches to Buckingham and Paddington, whilst making the Wendover branch navigable had cost an additional £13,000. Land purchase and rents on the Paddington arm alone had cost £11,900, and the wharves and other facilities at the basin a further £16,000.

Varying the line between Southall and West Drayton, through Cassiobury and Grove parks and across the Ouse valley, had cost £47,000 more than the estimate. Extra reservoirs had been provided costing £14,360. The canal had been made 6in deeper and slightly wider than planned, and the locks 6ft longer and 6in deeper, so as to admit the same size of boat as then operating on the river Trent. This had cost £16,740.

The expense of interest on capital, the cost of the wharf at Whitefriars, and the expense of dealing with the unforeseen problems such as at Braunston and Blisworth merely added to a long list totalling £640,000.

At the time it was calculated a further £185,000 would be needed to complete the works. These were Blisworth tunnel, Wolverton aqueduct, reservoirs on the Braunston and Tring summits, the Aylesbury railroad, and the Paddington water works. To tide matters over the shareholders agreed to £20,390 being raised by loan, this being the balance still left under existing powers. They also agreed on yet another approach to Parliament, the ninth, to authorise the raising of more money. The expenses of the first eight Acts alone had totalled £16,000.

A revised estimate was presented to the shareholders at the November meeting which showed at least £347,000 was needed. The main items were the tunnel, deep cuttings and locks at Blisworth and Stoke Bruerne costing £70,000; the aqueduct and embankment at Wolverton £40,000; water pipes and other works at Paddington £55,000; the proposed Boulters lock canal £55,000; and additional reservoirs £22,000.

In addition some land had still not been paid for; there were claims for damages that had to be met; and there were outstanding debts that needed to be settled. Extra facilities were needed at Paddington which would cost £10,000, and if the proposed branch was extended beyond Boulters lock to Reading another £80,000 would be needed. This brought the capital requirements up to almost £440,000.

The new Act[6] obtained in 1803 authorised raising another

A horsedrawn narrow boat leaving Fenny Stratford lock. This lock, built in 1802, has the smallest rise of any on the Grand Junction system, a mere 12 inches. *Author's collection*

£400,000. Much of the money was to be provided by a one-for-four rights issue to the existing share and loan holders. If fully subscribed this would yield £234,500. As much of the money was needed at Blisworth and Wolverton the Act authorised a toll of 3½p per ton through Blisworth tunnel and 1½p over Wolverton for coal, coke, lime, limestone, building materials, ironstone and manure. For all other goods it was double.

By June 1804 over £200,000 had been raised by the new subscription, whilst most of the first loan had been converted into shares. The company's capital stood at £1,328,295, making it probably the largest in the world[7].

Original whole shares of £100 each	£465,700
4,500 half shares of £50 each	225,000
First Optional Loan converted	163,888
New subscription	208,025
	1,062,613
First Optional Loan unconverted	1,112
Second Optional Loan	100,000
Third Optional Loan	28,000
Fourth Optional Loan	89,250
Mortgages outstanding	47,320
	£1,328,295

Even this was not the end, for in another Act[8], the tenth in the company's history, an additional toll was authorised on all goods except coal, coke, timber, manure, ashes, bricks, stone, limestone, ironstone, wrought iron and cast iron. A levy of 2½p per ton was also made on all goods except lime, ashes and manure which travelled less than eight miles on the canal, but passed through a lock. At the same time the clause in the first Act exempting all Government goods from toll was rescinded, as this had become a considerable burden to the company.

One of the original milestones erected by the Grand Junction Canal Company

Meanwhile receipts were rising. The first section of the canal had been opened at the end of 1794, but it was some years before the revenue built up to substantial levels. In 1795 a mere £377 was collected, but thereafter this increased rapidly as further stretches were opened and trade became established. By 1799 the revenue was almost £10,000. It exceeded £50,000 in 1805, £100,000 in 1808, £150,000 in 1813, to peak at £198,086 in 1836.

In May 1801 it was announced that toll and other income from September 1798 to March 1801 had produced £41,425 whilst operating expenses for the same period were £28,000. A maiden dividend of 2% was declared in the form of a credit on the shares until the company could afford to pay over the money, which it did in February 1803.

Receipts to December 1802 exceeded expenses by £10,572, and

a second dividend of $1\frac{1}{2}\%$ was declared in February 1803. This was paid in April. The capital stood at £694,925 at the time and the dividend absorbed £10,424.

A third dividend was declared in October 1805, and from 1806 dividends settled down to a regular pattern with one in June and the other in November. They were paid in August and the following January. Thereafter two payments were made each year except in 1816, when there was heavy expenditure on new reservoirs and other expenses.

One of the main traffics on the canal was coal. At the southern end of the canal coal was called "Sea Coal" because it was shipped by sea from north-east coast ports to London.

As early as 1667 this coal had been subjected to a toll to the Corporation of London to enable some of the damage wrought by the Great Fire of 1666 to be repaired. Later the money raised by the coal dues was used to carry out improvements such as the Victoria and Albert Embankments.

The Grand Junction posed a real threat to the established sea coal trade, for the canal was to have excellent connections with collieries in Warwickshire, Leicestershire and Nottinghamshire. Opposition from these shipping interests led to a clause in the company's original Act prohibiting the carriage of coal any further south on the canal than the entrance to the proposed Langleybury tunnel. The point was amended to the north-east end of Grove park when the deviated canal line was authorised.

In 1805 this total ban was eased and 50,000 tons were allowed to pass to Paddington upon payment of the coal dues. A man was stationed at Grove park to collect the money and keep a record of the tonnage passing this point. Later even this restriction was abolished and any amount of coal could pass subject to the payment of the dues. In 1831 these were standardised at $5\frac{1}{2}$p per ton.

One effect of the Coal Duties Acts was the establishment of a thriving business at Lady Capel's wharf at Grove, beyond where the restriction applied. With the ban on coal passing further southwards by canal it was offloaded here on to carts to continue its journey. A substantial trade was built up serving Watford and the immediate district round about. For many years the company leased this wharf from the Earl of Clarendon, but gave it up in 1847 when its tenant left.

A London coal duties obelisk.

Even with the restrictions the coal tonnages on the canal were soon substantial, 26,700 tons of sea coal being carried northwards from Brentford in 1805, with 94,730 tons of inland coal brought southwards from Braunston. By 1830 these figures had grown to 39,800 and 149,000 tons respectively.

The canal's early days were not solely involved with financial problems. One of the first difficulties arose at Fenny Stratford. It is sometimes assumed that the lock there, which has the smallest lift of any on the Grand Junction—a mere 12in—was due to a mistake in levels. This is not so.

Between Fenny and Wolverton the canal could not be made to hold water properly. Jessop attributed this to the rocky ground through which the canal had been cut and the lack of suitable material to line the bed. The two-mile stretch north of Fenny was worst as it was virtually impossible to maintain the water level high enough for fully loaded boats[9].

Matters became so bad that the canal had to be closed at this point in May 1802 for repairs to be carried out. To overcome the problem, and purely as a temporary measure, a lock was built at Fenny in June. The level of the entire eleven-and-a-half-mile pound to the top lock of the flight down into the Ouse valley at Wolverton was then lowered by about 1ft[10] to below the point where most of the leaks were occurring. It was hoped the leaks could be cured and the level restored, thus making the lock unnecessary.

An estimate of this work was made in October 1805, but no action was taken due to the costs involved and the delays to traffic whilst the work was being carried out. Matters were reconsidered in 1838 with a suggestion to raise the long Fenny pound by 6in and deepen the three-mile pound southwards to Stoke Hammond. Again the expense was a deterrent, and the lock remains today as a permanent part of the Grand Junction scene.

Another problem arose near Hemel Hempstead. As the canal descends from the Tring summit it runs parallel with the Bulbourne, and there are several links between the two. For instance at Winkwell they unite for a short distance and divide again. The river then flows through the Boxmoor watercress beds, passes under the canal near the Fishery Inn through a three-arched culvert and joins the river Gade just above Two Waters mill, whilst the canal passes down through the Boxmoor locks to Hemel Hempstead.

The Gade, which rises to the north west of Great Gaddesden, crosses the canal on the level at a right-angled junction a short distance from its confluence with the Bulbourne. Hence the canal below Hemel could be fed by the waters of both rivers, and this led to difficulties.

Downstream from the Gade junction the canal used to run for over one and a half miles without a lock. This was known as the "long pound". There was then a flight of four locks to lower the canal 28ft down to the Kings Langley pound. This arrangement was criticised bitterly by the owners of Apsley and Nash mills, who contended that much of the river water was being diverted to supply the long pound and the flight of locks, rather than their mills.

To counteract these claims the company decided in July 1805 to erect a steam engine which would pump water from the bottom of the four locks back up to the long pound. The engine was supplied by Boulton & Watt of Birmingham and erected not far from Nash Mill.

One of the millers was George Stafford of Apsley mill. In

February 1807 he claimed £124 for the inability to work his mill between June and December 1805 due to the diversion of part of the water supply. In May 1809 John Dickinson purchased the mill from Stafford[11], and early in 1811 he bought Nash mill as well. Dickinson's partner was George Longman, who provided much of the finance; the firm traded as Longman & Dickinson. Both Apsley and Nash had been mentioned in the Domesday Book as corn mills, but both had been converted to papermaking towards the end of the eighteenth century[12].

The steam engine at Nash did not solve the problems. In August 1811 Dickinson offered to accept £12,000 for his claims. The company then suggested paying a fixed sum in compensation for the next three years instead of working the engine. Its annual costs were averaging £500, which the company thought much too high. The long pound caused much of the trouble, since it was not watertight, but Bevan thought the leaks could be cured in a few years, after which the engine should not be needed.

Discussions with Dickinson failed to reach agreement, and in July 1812 Longman & Dickinson sued the company for damages. The action was heard before the Master of the Rolls and, whilst the millers' claim was disallowed, the Court upheld some of their grievances and ordered the company to prevent any unnecessary waste of water in the long pound.

Dickinson made new proposals in February 1813, suggesting that a larger steam engine should be erected and worked by the company to pump back all the water being drawn from the river by the canal. As an alternative the existing engine should be transferred to him, with the company supplying enough coal to operate it for seven months a year, and reimbursing labour and maintenance costs. Bevan examined these proposals but rejected them.

Negotiations again having failed, the millers brought another action claiming £4,463 damages for water diverted between May 1814 and January 1815. A sub-committee was appointed to handle the situation and Thomas Telford was called in to advise the company. He carried out a series of tests over three days and concluded the steam engine more than compensated for any water lost by navigation, by the differing capacity of the locks, by leakage or any other causes claimed by the millers.

A few days earlier the millers' engineer Donkin had also conducted tests, and his results purported to show the canal was diverting water from the rivers. He claimed no less than one-fifth of the total flow of the rivers was being taken, even after allowing for the water returned by the engine.

The new action was heard before a special jury at Hertford Assizes on 3 March and the assessment of damages was referred to Mr Sergeant Bosanquet, who had power to conduct further enquiries during the summer.

The company now decided to build side ponds at the four locks to cut down on their water consumption. Bevan was ordered to start work on them in July, but there was some trouble in obtaining the land needed as it was owned by Dickinson, who was not co-operative. The company was within its Parliamentary powers and precautions were taken to ensure there was no interference with the work.

The side ponds were brought into use but the millers refused to accept them as a solution to the dispute. They considered the company should have built a reservoir that was stipulated in the original Act to supply the mills on the Gade and Bulbourne.

The dispute lingered on, and in February 1816 the company decided to suffer judgement by default so that the whole matter could be tried by the sheriff. This led to a heavy award in favour of the millers for the water abstracted between 1809 and 1815. As this was unexpected it was a major contributory reason for the company paying no dividend in June[13], and only 2% in December. This was not the end, for Longman & Dickinson started another action in June for damages for subsequent abstractions.

In January 1817 the first steps were taken towards seeking a permanent solution, and in February Dickinson produced proposals for a deviation of the canal. There was a meeting in April following which Bevan was asked to report on the costs of puddling and maintaining the long pound, of operating and maintaining the four locks and the side ponds, and of operating and maintaining the steam engine.

Meanwhile Telford was asked to survey the proposed deviation and to report on the idea. He was also asked whether the steam engine and the side ponds together would be sufficient to protect the company against any further claims, provided the long pound could be made watertight.

Following Telford's survey, which he made with his assistant William Provis, the company adopted the idea of a diverted canal as the best way of settling the dispute[14]. An agreement was reached with Longman & Dickinson and a Bill was prepared for submission to Parliament. In December 1817 negotiations for the necessary land were started with the landowners.

The Act[15] was passed in 1818. It authorised the abandonment of the canal from Frogmore swing bridge to the junction with the tail

The canal diversion past Apsley and Nash mills.

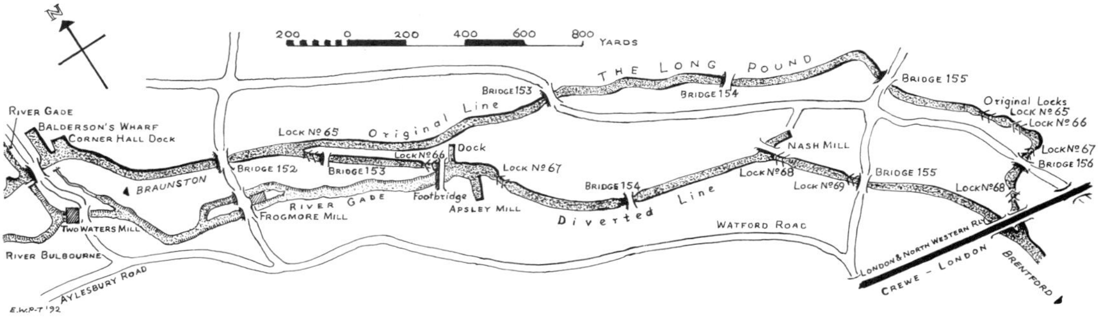

A southbound narrow boat waiting for Apsley top lock to fill. Immediately above this lock the line of the original canal can be seen to the right. *National Waterways Museum*

of Nash mill and building a new line along the course of the river Gade. The company could borrow £30,000 for the project.

Work started at once, with the new locks being built by Dickinson under contract. Early in August he was paid £3,000 for work already completed. By February 1819 the diversion was nearly complete, and it was opened soon afterwards. The total of Dickinson's contract was £10,580, which included £600 for the old bricks, somewhat less than Telford's estimate of £12,000. The company carried out the earthworks.

The new canal settled the dispute with the millers. Five new locks with a lesser fall, and hence using less water, were provided to replace the four old locks, and much of the long pound was bypassed. To avoid renumbering all the locks southwards lock 69 at Kings Langley was renumbered 69A.

The water problem at Apsley and Nash mills was eased soon afterwards when both were converted to steam working. The coal was supplied by canal but large quantities of water were still needed in the papermaking process.

In 1826 Dickinson opened Home Park mill, just over a mile downstream from Nash, and in 1830 Croxley mill, south of Watford, was opened[16]. Both depended on water power in their early days.

There were problems with millers in other places. It was almost inevitable that whenever the company needed to alter levels to improve conditions for navigation there would be repercussions from millers both up and downstream. The canal was a mixed blessing to millers, for whilst they welcomed the advantages of cheap transport they were more concerned to safeguard their water supplies. In the original Act several clauses were inserted to protect their interests.

The problems at Strutts' Batchworth mill near Rickmansworth were typical. Barnes reported on raising the level here in 1802, but this was still being disputed in March 1809 when the owners offered to sell. In June they asked for £17,000 against the company's offer of £13,500. By December the company had increased its offer to £16,000 and an agreement was then reached.

Despite this it was not until May 1823 that the water level from Lot Mead to Batchworth mill was finally fixed[17], due to the many factors involved in this vital stretch of canal joined by both the rivers Gade and Colne.

At Brentford there was another problem with the millers. Originally there was no lock downstream from Brentford lock No 100 and Brentford bridge, the Brent being tidal to this point. Boats could only pass for limited periods at high tide. In September 1802 several canal users complained about the general delays and inconvenience of navigating this section. They suggested either widening and deepening the river or cutting a new channel southwards to the deep waters of the Thames.

To try and prolong the brief period of time when navigation was possible the company erected a set of flood gates which held up the level for a while after the tide had turned. These brought complaints from Richard Kidd, whose mill was on this part of the river, about the water backing up and preventing his mill wheel working properly. Kidd asked either for an allowance towards the

John Dickinson & Company used the canal for many of their transport needs. The steamer *Countess* operated an express service taking finished paper products down to the company's Paddington depot, returning with rags and other raw materials for the papermaking process. This picture was taken about 1901 at Apsley South End Wharf. *John Porter*

cost of altering his wheel, with a £100 annuity for the privilege of holding up the water, or that his interest should be bought out for £2,000. William Jackson, the company's accountant, negotiated with Kidd and succeeded in acquiring the mill rights for £1,500. In 1818 the mill was demolished and a set of sluices erected on its site.

The obvious course would have been to build a lock at the entrance to the canal in the first place. The company's Act, however, specified that the free navigation of the Brent was not to be impeded in any way. Once the company owned the mill rights it was easier to convince the landowners and others that it would be in everyone's best interests to have a proper lock. Thames lock was then constructed 250 yards from the Thames junction in place of the flood gates.

Once the main canal was complete the problems confronting the company tended to diminish and it could concentrate on more positive matters.

As traffic increased it became necessary to appoint toll clerks to collect the tonnage dues at various key points. The first two had been stationed at Brentford and Uxbridge and a third was appointed at Braunston in 1796. By January 1802 there were eleven in the company's employ.

The senior, and highest paid, was James Cherry at Braunston, on a £90 salary, showing that this was already the most important

A view inside John Dickinson's Apsley South End Wharf showing narrow boats loaded variously with paper products and with bundles of rags. *John Dickinson & Co Ltd*

station on the canal. He was the first of a long line of Cherrys in the Grand Junction's service. The clerk at Brentford was paid £85, Paddington £80, Blisworth £75, Stoke Bruerne £70, Marsworth £60, Cosgrove £55, with Boxmoor, Bulls Bridge, Great Linford and Harefield all at £50. In 1805 an assistant was appointed at Braunston to help Cherry, with a second in September 1808, indicating Braunston's continuing role as the leading toll station.

As trade grew and new sections of canal opened changes were made to these toll stations. When Blisworth tunnel was complete the station at Stoke was closed, and with the opening of the Northampton branch the Blisworth station was moved to the junction. Likewise when the Grand Union Canal opened a station was established at Long Buckby. In 1818 an office was opened at Fenny Stratford, but this had a short life and was moved to Leighton Buzzard.

Under the company's Act all boats were supposed to have graduated metal scales fixed to their sides so that the weight of the cargo could be determined. To assist the toll clerks the company decided in August 1799 that all boats should be gauged and their displacement recorded for known loadings. These records were kept at all the toll stations and the clerks, by measuring displacement,

FMC's steamer *Phoenix* being gauged for toll purposes about 1910 in Long Buckby Lock (No. 9). In 1916 the gauging point was moved to Norton, at the junction with the Leicester line. *National Waterways Museum*

could calculate the tonnage from the gauging table. The gauging of boats was started at Paddington in 1802[18] and soon became standard throughout the Grand Junction system. A second gauging station was opened at Braunston in 1816.

In June 1805 the company's engineer James Barnes retired after thirteen years' untiring service. He had dedicated himself to the Grand Junction, and its completion represented his crowning achievement. Even in retirement he was not far away, as he lived at Braunston.

In June 1803 the canal had been divided into three districts for administrative purposes. Initially these were the Northern district from Braunston to Leighton Buzzard, the Middle district from Leighton to Hunton Bridge, and the Southern from Hunton Bridge to Brentford and Paddington.

John Woodhouse was appointed engineer of the Northern district in September 1805. He was succeeded after a few years by William Thompson, who remained until his retirement in 1835. When Barnes retired Benjamin Bevan and Henry Provis looked after the Middle and Southern districts on an informal basis. Bevan left the company at the end of April 1817 and Thomas Lake was appointed in May to replace him. Provis resigned in 1816 and Richard Trubshaw was appointed in November, but he was dismissed in December 1821, when John Holland took his place.

Philip Box remained as the company's treasurer until November 1801 when he entered into a partnership with William Praed, the company's chairman. Box and Praed were then appointed joint treasurers.

In the early days the company often had to rely on financial help from its treasurers, and in November 1802 it was decided to merge the company's two offices in London near to the treasurers' offices in Fleet Street. In January 1803 the offices were moved to Essex Street, Strand, but in 1805 another move was made to 21 Surrey Street, Strand.

Acton Chaplin and Edward Gray resigned as clerks to the company in October 1805 and Charles Harvey, appointed superintendent in June, stepped in to act temporarily. In November Richard Cowlishaw Sale, the committee clerk, was appointed clerk, with Harvey becoming secretary.

Harvey was replaced by Charles Simpson of Lichfield in March 1816. Simpson died in December 1820, and the office of secretary was then discontinued and the decision taken to have a permanent chairman.

William Praed stepped down after serving for nearly thirty years and the Hon Philip Pleydell Bouverie was appointed chairman in February 1821. His role was equivalent to that of a general manager; he received a £500 salary and £100 expenses allowance. From the same day Sale took on the combined duties of solicitor and clerk.

In 1816 the shareholders agreed that a Select Committee should be set up to assist the general committee. It was to be a

The monument to William Praed, the company's first chairman, in Tyringham Church, near Newport Pagnell. *Alan Faulkner*

smaller and more streamlined body, with a maximum of nine members. Praed was its chairman, with Thomas Cobb his deputy. Its first meeting was on 11 June 1816; its first resolution was that a clock should be bought for the committee room. Thereafter meetings were held at regular intervals and the Select Committee conducted most of the day-to-day business that would otherwise have gone to the full committee.

In 1803 a sinking fund had been set up. By October 1808 this totalled £70,000[19]. The fund was in such a healthy state that in October 1817 it was decided to repay the various loan stocks, the Aylesbury canal subscription and a mortgage on land at Wolverton. Most of the money was invested in Government exchequer bills which attracted interest at only 3%, whereas the company was paying 5% on its borrowings.

In January 1818 Sale gave formal notice to pay off all the mortgages and loans in September, when interest on them would stop. Those holding over £200 could transfer to a new loan stock carrying interest at 4%.

The largest new borrowing incurred since 1802 had been a fifth optional loan issued in November 1804, which had attracted £75,513 by March 1806. At the time of the loan repayments in 1818 the issued capital stood at £1,181,440. It remained at this figure until 1820, when a gradual reduction started. By 1845 this had reached £1,145,000. With the regular and increasing dividends the financial position had improved very considerably since the early 1800s.

NOTES TO CHAPTER FIVE

1 33 Geo III, c 80—30 April 1793.
2 36 Geo III, c 25—24 December 1795.
3 38 Geo III, c 33—26 May 1798.
4 41 Geo III, c 71—20 June 1801.
5 Report to General Assembly, 8 June 1802.
6 43 Geo III, c 8—24 March 1803.
7 Grand Junction printed circular, 4 June 1804.
8 45 Geo III, c 68—27 June 1805.
9 Report to General Assembly, 8 June 1802.
10 Ibid.
11 Evans, Joan. *The Endless Web,* 1955.
12 Ibid.
13 Grand Junction printed circular, 13 June 1816.
14 General Committee minutes, 8 May 1817.
15 58 Geo III, c 16—17 March 1818.
16 *The Endless Web,* op cit.
17 Inscription on monument at Rickmansworth Lodge, near Batchworth lock.
18 The Grand Junction gauging registers start from 5 July 1802.
19 *Northampton Mercury,* 12 November 1808.

Reservoirs and Water Supplies 6

ONE OF THE biggest problems confronting the Grand Junction was to provide an adequate supply of water to its two summit levels. Both presented difficulties, but those at Tring were more acute than those at Braunston.

A few small streams were tapped on the three-mile Tring summit, but they were of only minor significance. The main supply has always been from the Wendover feeder branch, which joins the northern end of the summit just above Marsworth top lock and runs along the north side of the Chilterns, intercepting several streams on its course to Wendover. These occur at the junction of the various strata of which the hills are made up—a water-bearing chalk lying on a thin layer of Totternhoe stone which lies on an impervious chalk marl.

From Marsworth the first is the Tring feeder, which enters north west of New Mill. Originally this was fed by the surplus water not required to operate the mill. After being purchased by the company the mill went out of use and the entire supply, varying from eight to twenty-five locks[1] daily, now enters the canal[2].

Another stream enters about half a mile from the Wendover end of the branch. In 1848 nearly £2,000 was spent on culverting it and driving an underground tunnel, or heading, into the nearby hillside to increase its yield. This source, called "The Heading", yields from two to ten locks daily.

The main source, which enters at the extreme end of the branch, is the Wendover stream, a natural watercourse starting nearly a mile south of the canal and collecting water from various ponds on the way. It formerly ran into the river Thame and provides from four to twenty locks daily[3].

To compensate millers at Aylesbury for the stream water diverted into the canal a reservoir was built in 1798 a short distance from the branch at Weston Turville. It was supplied by two small streams and, in winter, by surplus water from the other feeders. Later the company purchased the milling rights and the practice of supplying compensation water was discontinued.

In 1814, when demand for water was particularly high, a well and heading were built so that water from the reservoir could be pumped into the branch by a steam engine. Whilst intended as a temporary measure, the engine remained in use for nearly twenty-five years.

Once the canal was opened over Tring summit it was found that in summer water supplies were barely adequate, a shortcoming soon highlighted during a drought. To ease the problem construction of a new reservoir was started in 1802 on low-lying ground near

Wilstone, to the north of the branch. It was fed by two small springs, the Ashwell Head and the Barwell Head, and by a stream from Drayton Beauchamp, the Drayton feeder. Arrangements were also made to run surplus water off the summit to the new reservoir by means of a link at Drayton connecting with the Drayton feeder.

To draw water from Wilstone a well was sunk beside the branch at Whitehouses and a heading was built to the reservoir. A thirty-lock-per-day capacity pump was installed at Whitehouses with a steam engine to drive it. Major difficulties were experienced in driving the 700yd underground heading, as large quantities of water were encountered, and the work was not completed until June 1803[4].

In the spring of 1802 leakage through the banks of the branch had been noticed. Temporary repairs were made, but in the autumn of 1803 the branch between Whitehouses and Marsworth had to be closed for more comprehensive work. It was not reopened until early summer 1804, and even then the repairs were incomplete. During the closure all the Wendover water was run via the Drayton feeder down into Wilstone, and once this was full it went to waste.

The Tring group of reservoirs and feeders.

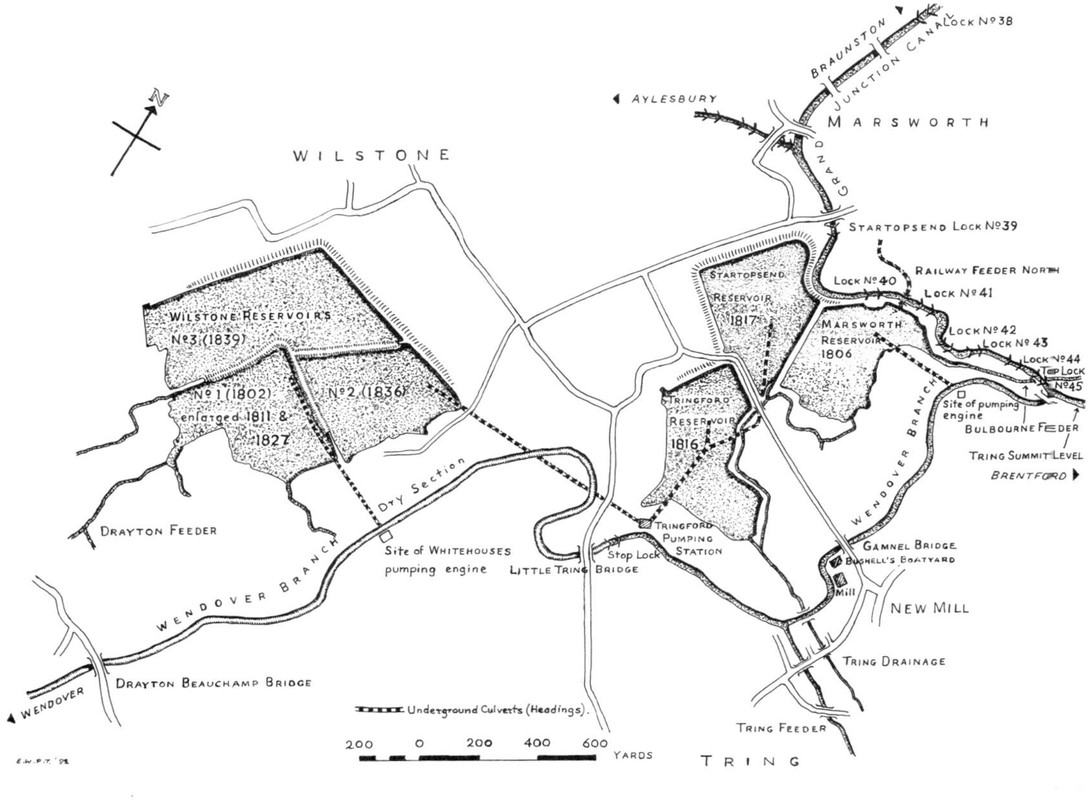

Benjamin Bevan, who was in charge of the repairs from March 1804, proposed another reservoir to store this surplus water from the summit. His plan was adopted in December 1805 and the reservoir built in 1806 between the branch and the Marsworth flight of locks. It was supplied by running water down the locks from the summit and through links above Marsworth locks 41 and 40. Water was supplied back to the canal at Startopsend lock 39 in Marsworth.

The new reservoir, called Marsworth, also has an independent supply from a spring rising near Bulbourne, north of Tring. The Bulbourne feeder has now been culverted throughout its length. It passes under the main canal and the branch on its way and yields from one to seven locks daily.

In 1810 a well was sunk alongside the branch not far from the main line junction, a steam pump set up and a 350yd heading driven to enable water from Marsworth to be pumped direct to the summit in times of shortage.

The company's 1812 Act[5] authorised an additional reservoir on the summit to help supply the Aylesbury branch. In fact two reservoirs were built. The first, a high level one named Tringford, was started in April 1814 and completed in 1816; the second, Startopsend alongside Marsworth reservoir, was begun in 1815 and finished in 1817.

Tringford is fed by the Tring Drainage, which passes under the Wendover branch in a culvert and is supplied by surface drainage from various parts of Tring[6]. Before New Mill was purchased by the company its water was discharged into the Tring Drainage after passing the mill wheel. It yields from one to fifteen locks daily.

A crowd of boats on the Tring summit level of the canal at Cowroast lock, probably in the early 1920s. *National Waterways Museum*

When Tringford is full the supply can be diverted into Startopsend, which is also fed by the overflow from the adjoining Marsworth reservoir.

After Startopsend was opened the pumping engine near the main line junction was dismantled, as surplus water from Marsworth could be run into Startopsend. The pumping well and heading were then used to run surplus water from the summit to these reservoirs instead.

Nowadays another source of supply for both Marsworth and Startopsend is from the Tring sewage works, situated between Marsworth reservoir and the branch; the supply produces about four locks daily.

With the new reservoirs came a fourth pumping station on the branch, at Tringford, just over a mile from the junction, on a site selected by Thomas Telford. It was destined to survive all the others and to become the central station for the reservoirs on the summit.

At first a 450yd heading was built from the station to Tringford reservoir, whilst a 500yd branch served Startopsend. These were completed in the summer of 1817[7]. Earlier in the year a steam engine had been ordered from Boulton & Watt. Meanwhile construction of the engine house went ahead and was completed in April 1818. Installation of the engine and pump then began and by the middle of August pumping had started. The engine cost £3,120.

In 1811, and again in 1827, the size of Wilstone reservoir was increased by raising its banks, bringing its capacity to over 1,400 locks. In May 1835 work started on a new reservoir along its east side[8], which became known as Wilstone No 2. It was completed in January 1836 and is the same size as Wilstone No 1. Water from No 2 can pass to No 1 and hence to the summit. It has no natural supplies of its own and is fed from No 1 and from the overflow from Tringford and Startopsend.

By the end of 1835 the steam engine at Whitehouses was worn out. John Holland proposed a new heading should be built from Wilstone No 2 to the Tringford pumping station[9], to allow the Whitehouses engine to be scrapped. As the new heading was to be lower than that from Whitehouses it would also allow more of the Wilstone water to be pumped out.

Holland's plan was adopted and the driving of the 950yd heading started early in 1836. Due to the magnitude of the task it took over one and a half years to complete, and even then a portion was left without a brick lining. Pumping from Wilstone to Tringford started on 21 November 1837. The entire pumping operations were now centralised at one point.

Following a report in February 1836 by William Anderson, a consulting engineer, modifications were carried out to the Tringford engine to enable it to pump from the greater depth now required. The lift from Wilstone varies from 60 to 75ft, depending on the level in the reservoir, compared with 35 to 50ft from Startopsend and 20 to 35ft from Tringford.

Tringford pumping station beside the Wendover branch, pictured here in September 1956. *Alan Faulkner*

The alterations included building a new and deeper well to connect with the heading from Wilstone, and re-erecting the engine to draw from this new shaft. The original well shaft, which was connected to the heading to Tringford and Startopsend, was then linked to the new shaft. The work cost nearly £2,000.

In October 1838 a tender was awarded to Hugh McIntosh to build yet another extension to Wilstone. This was almost as large as the other two together. Wilstone No 3, completed in 1839, can hold 2,600 locks[10]. It lies to the north of the others and is connected to No 2.

At the same time a new overflow weir to serve all three Wilstone reservoirs was built at the north-west corner of No 3 to enable water to be run into the nearby Gudgeon Brook. There is a connection from this brook to the Aylesbury branch, just below the ninth lock, allowing it to be supplied from Wilstone if necessary.

The old engine at Whitehouses was sold in April 1841, together with that formerly used at Weston Turville. The buildings at Whitehouses were converted into a dwelling house.

With the concentration of all pumping on one engine, problems could arise during a breakdown or normal repair work. In December 1837 the company acquired a second-hand pumping engine for £600 from Hunter & English, a well-known firm of engineers who had carried out some modifications on the Tringford engine. This new engine became known as the "York" engine and was installed in the original shaft, or "Old Pit", formerly occupied by the modified "Old" engine, which was now working in the "New Pit".

This was all done in the summer of 1838. At the same time preparations were made to deepen the Old Pit and make a separate link from it to the Wilstone heading to enable the York engine to draw from Wilstone. It proved impossible to carry out this work whilst the Old engine was operating, but in June 1839, during an enforced stoppage for repairs, the work was completed.

A year later it was found that part of the foundations of the Old engine had sunk, causing another stoppage. The engineer William Cubitt was called in to inspect the damage, and he found the brickwork had been crushed by the continual working of the engine. Remedial measures were then taken to counteract the concussion caused by a column of water, weighing fourteen tons, on the foundations at every stroke of the engine.

There was an accident to the York engine in November 1843 when valve nuts worked loose and considerable damage resulted. Repairs were not finished until January 1844. The incident revealed a serious defect in the pumping arrangements. As water from Startopsend and Tringford came into the Old pit under the York engine, repairs could not be carried out whilst the Old engine was drawing from these reservoirs. Similarly once the repairs had started the Old engine was unable to draw from the two reservoirs.

To overcome this a new 40yd long heading was driven to allow water from them to enter the New Pit, without having to pass through the Old Pit. It took four weeks to build, and the short connection between the two well shafts was then sealed off. Either engine could now work from either group of reservoirs independently.

During all this time the Wendover branch continued to leak. Spasmodic repairs were carried out but in 1841 it was found that twenty locks a day were being lost[11]. In 1844 Cubitt was called in to advise. He recommended trying to cure the worst leaks only, since relining the entire branch would cost £12,000. Extra puddle was then put in on both sides of the canal for four and a half miles at a cost of £2,654.

Before this was done the branch often had to be dammed off to protect the level in the main canal. The Wendover water was then run down into the reservoirs. The closures caused considerable hardship to the Wendover traders, who had to divert their goods to Aylesbury, with the company paying them compensation.

Ten years later, and despite the money spent in 1844, the branch was leaking as badly as ever. A test in 1855 showed a daily loss of twenty-five locks[12]. One proposal was to repuddle the branch throughout. This would cost £12,500 and some work was carried out.

An alternative was to line the canal bed with a 2½in thick layer of asphalt. Cubitt supported this plan, estimating the cost at £700 per mile, against £2,700 per mile for repuddling. The asphalting was carried out on the worst stretch of over 4,800yd between Little Tring and Aston Clinton. It was completed in 1858 and cost £6,465[13].

The relief was only temporary. In 1860 further repairs were needed, but in 1870 the leakage was worse than ever at almost thirty locks daily. Further repairs to the asphalt and the puddle were carried out. In 1871 tests showed ten locks per day were being lost from Little Tring to Aston Clinton and nine locks per day from Aston Clinton to Wendover.

These tests also showed that the asphalt had cracked in places and had been damaged by boats rubbing against it. As a result it was removed from the worst sections and the puddle restored. The company charged £2,000 against reserves for this special work. Another major effort was made in 1895 and 1896 when £1,650 was spent on remedial measures.

Until 1838 the only significant water supply feeding directly into the summit, apart from that from Wendover, was from Bulbourne Head where a spring fed into the canal half a mile south of Marsworth top lock. During the building of the London & Birmingham Railway (L&BR) through the Chilterns near Tring this spring was diverted so that it no longer fed the canal.

The company began legal proceedings against the L&BR in March 1839. This action was suspended in 1841 to allow the railway to build a culvert from just south of Tring station into the summit level about 200yd south of Marshcroft bridge—the high level turnover bridge in the middle of the cutting. The culvert, known as the Railway Feeder South, drains the southern end of the three-mile railway cutting. It produces about five locks a day, although in summer it usually dries up.

The northern part of the railway cutting is drained by another culvert, the Railway Feeder North, which was constructed by the Grand Junction in 1838. Up to thirty locks per day have been obtained from it.

In May 1843 the dispute with the L&BR was settled, with the railway paying £5,200 in compensation and transferring to the Grand Junction the Railway Feeder South together with the right to the water flowing through both this and the northern feeder.

The water in this northern feeder can be diverted into the Mottimore brook, which flows into the canal below Marsworth bottom lock 37. Part way along the feeder there is another connection with the canal above lock 38. The feeder then discharges into the canal between Marsworth locks 40 and 41, immediately opposite the weir into Marsworth reservoir. The Railway Feeder North is now the main independent supply to the reservoir.

In 1838 William Anderson recommended the provision of a series of engines to pump some of the lockage water from the Ouse valley to the reservoirs. A contract for nearly £13,000 was then awarded to Grissell & Peto to build the engine houses and culverts. By 1841 nine pumps had been installed to pass the water round the seventeen locks involved. They were at Fenny Stratford (pumping round lock 22); Stoke Hammond (23); Soulbury (24 to 26); Leighton Buzzard (27); Church (28 & 29); Horton (30 & 31); Ivinghoe (32 & 33); Seabrook (34 to 36); and Marsworth (37 & 38). The engine houses were built of brick with slate roofs. The pumps were powered by steam engines and were known as the Northern Engines.

Originally the Marsworth engine pumped into the Railway Feeder North and hence into Marsworth reservoir. From there water could be pumped to the summit by the Tringford engine. In practice it was found that the feeder was too small to cope with the increased flow, and the system was modified for the water to enter the canal above lock 38. From this pound it could be run into Startopsend reservoir.

The Northern Engines also pumped from several streams which feed the canal north of the summit. The main supply is from Ledburn brook, which enters below Grove lock 28. Other streams are the Mottimore brook, the Pitstone brook entering below Seabrook lock 35, a small stream immediately below the Soulbury three locks, another immediately above Stoke Hammond lock, and a further one half a mile below this lock[14].

As early as 1848 the Grand Junction had started what was to prove a lengthy search for water to be drawn from wells. The first were sunk alongside Cowroast lock and just south at Dudswell. The Dudswell well was a failure, but that at Cowroast became an important source of water. Its sinking brought the company into another conflict with John Dickinson.

This followed several dry summers when many springs dried up completely. Dickinson claimed this was due to the company pumping from the well, and he brought an action contending it had no right to well water. In January 1852 judgement was given against the company[15].

Some years later, during another dry spell, the springs dried up just the same. The millers, who were desperate for water, approached the company which found it was able to pump from its well, thus proving the water came from another source. The injunction was then set aside by mutual consent.

Between 1859 and 1871 a well was sunk in Wendover 100yd north of the church and a portable steam engine was installed to pump into the nearby Wendover stream. This provided nine locks a day. A permanent engine was considered but then it was found the well would deliver two locks daily without any pumping. Pipes were then laid to feed the water the 800yd to the branch by gravity, the pipes later being replaced by a heading.

In 1875 a second well was sunk in Wendover near the Hale Road. Its yield was unsatisfactory and it was abandoned in 1876, but was reinstated in 1891 when a heading was built to link up with that from the other well. The second now yields from one to four locks daily[16].

Loaded boats tied up at Marsworth in the summer of 1955, just below Startopsend lock. This section of the canal also had to supply the Aylesbury branch. *Alan Faulkner*

Between 1859 and 1862 a trial boring was made in the yard of the company's workshops which had been established at Bulbourne. This eventually reached a depth of 500ft but was unsuccessful. In 1896 another well was started at nearby Bulbourne Head, on the site of the dried-up spring. By 1900 the well was 80ft deep with a boring of another 146ft from the bottom. The yield was poor and the work was abandoned. A 1902 plan to deepen it to 500ft was not carried out.

In 1855 work started on sinking a well at Ivinghoe, a beam engine and pump being installed to deliver the water to the canal. This well was successful, and two more were sunk in 1864 and connected to the original so that the pump could draw from all three. Later problems were encountered with the wells filling with sand. Several attempts were made to clear them in the early 1900s but they were only partially successful and the yield gradually declined.

Braunston Summit

The problems on the Braunston summit were never as complicated as those at Tring. Originally two reservoirs were built and several feeder streams tapped.

Braunston reservoir was small—just over 300yd long and 70yd wide, with a water area of just over five acres—and lay immediately to the south of the canal midway between the top lock and the north-west end of the tunnel. As it had a restricted catchment area, being fed by one small stream, it was not very successful.

The second reservoir, known as Drayton, or Daventry Old, reservoir, lies along the west side of the Daventry to Kilsby road just south of the line of the tunnel. It has an area of twenty-seven acres and a capacity of 1,300 locks when full, and is fed by a stream from the west with a catchment area of over 500 acres. Its outlet into the summit is alongside the south eastern portal of the tunnel.

Another stream feeds into the northern tip of the reservoir. This has been culverted for most of its length from the north and can only discharge into Drayton when the level is several feet down; when the reservoir is full the stream can be diverted to feed directly into the summit.

Following reports by Jessop and Barnes in August 1795 work on

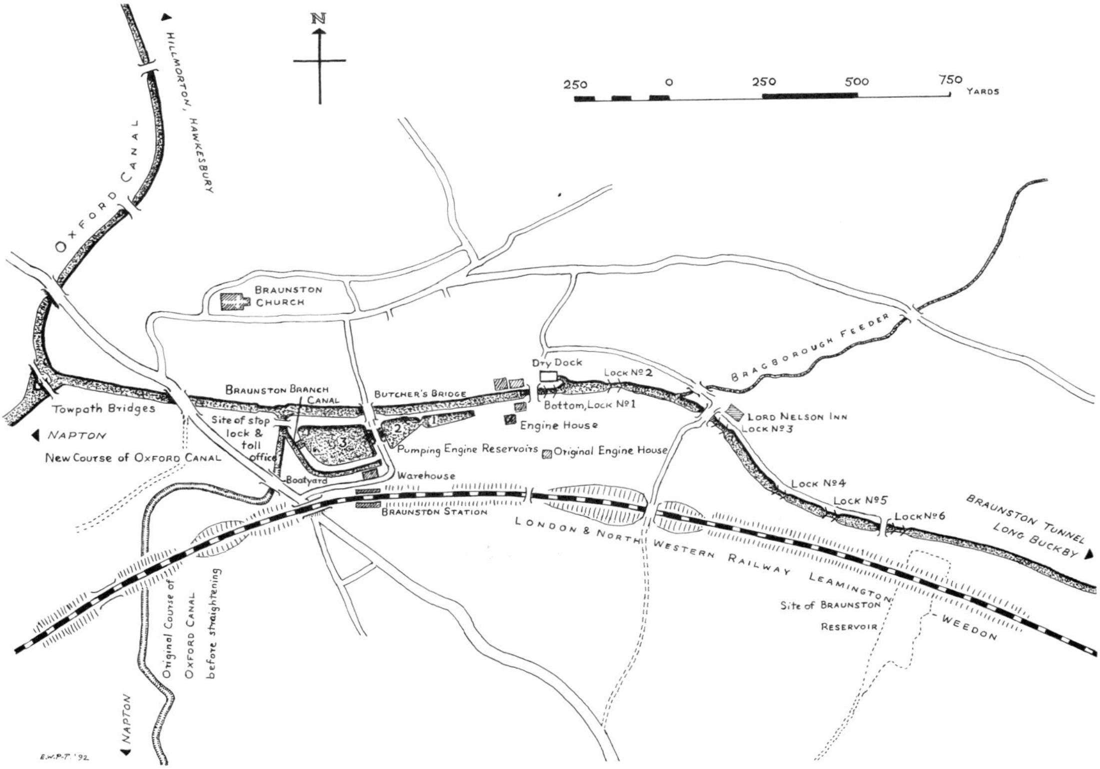

The canal at Braunston.

both Braunston and Drayton was started at the end of that year. They were completed in time for the opening of the canal across the summit in June 1796.

With the rapid increase in traffic, even before the main line was finally completed, the two reservoirs were soon shown to be inadequate. A third, and larger, reservoir was then built on part of the line of the proposed Daventry branch canal. Known as Daventry, or Daventry New, reservoir, it lies to the east of the Daventry to Welton road half a mile south of the canal.

In March 1803 Barker started negotiating for the seventy-five acres of land needed. Some of this was on Daventry common and its acquisition was delayed pending the passing of an Act enclosing land around the town, so Barnes was unable to start construction until the summer of 1803[17]. In August an extension was authorised involving the purchase of extra land.

The reservoir was in use by the middle of 1804, although some work still needed completing. It cost about £20,000, of which over £9,000 was for the land. Daventry originally could hold 6,470 locks when full[18], has an area of almost 100 acres and is fed by four streams running down the valley.

The two main streams, which enter at the south-west end, drain much of Daventry. The third enters immediately by the headbank. The fourth, which enters on the eastern side, used to tap several springs to the east of Daventry. Later some of these springs were used to provide drinking water supplies and little now enters the reservoir from this source. The total catchment area is over 2,100 acres and the outlet feeds into the summit just over half a mile from the south-east end of the tunnel.

Soon after Daventry reservoir came into service Braunston was drained and the feeder stream culverted through the headbank to enter the canal below water level. In April 1893 the site was purchased for the Weedon to Leamington railway line and an embankment was built across the reservoir parallel to, and only a few yards from, the old headbank, with the stream being culverted under the railway as well.

In September 1796 Barnes had been asked to survey a proposed reservoir and feeder from Watford to the main canal. He recommended a reservoir in the parishes of Welton and Ashby and a lengthy feeder through Watford. No action was taken on the reservoir, but the feeder became an important source of supply to the summit.

At the south end of the Leicester line (see Chapter 8) today there are three feeders. The Crick feeder enters the Leicester summit just south of Crick tunnel; the Watford feeder enters from the west just below Watford locks; and the Old feeder enters, again from the west, just south of the Watling Street bridge. Then the Welton feeder enters the Grand Junction's summit from the north just over 500yd east of Welton wharf.

All four were originally part of the Watford feeder which entered the summit where the Welton feeder now comes in. As built it took a roundabout course and by following the contour was virtually all on one level. It intercepted several streams on the way, including the Crick water from its higher level at the northern end. There was a tunnel taking the feeder under high ground at Weltonfield farm.

When the Grand Union Canal was built in 1814 its the line crossed the Watford feeder twice, and the severed sections were then altered so as to flow into the new canal as a series of independent feeders. The Old feeder had its flow reversed for part of its length, whilst the Crick feeder was diverted to run directly into the Grand Union's summit. Parts of the line were abandoned, with the land reverting to the original owners.

The Watford feeder today has a catchment area of almost 2,800 acres and the Old feeder over 700 acres. In summer only a small supply is provided, but in winter the feeders contribute a significant quantity of water.

Even with the new Daventry reservoir in use there were still shortages, and in April 1805 authority was given to erect a steam engine at the bottom of Braunston locks to pump water back to the

To prevent any unnecessary waste of water it was essential that the locks were kept in good repair. Here work is in hand on one of the Braunston locks in June 1905. *National Waterways Museum*

summit. This was brought into service in the middle of June, having been transferred from Blisworth where it had been used on the tunnel workings. It pumped from two small reservoirs provided alongside the canal at Braunston between the bottom lock and the junction with the Oxford canal. A channel led from the engine to discharge into the summit just above the top lock.

These reservoirs were supplied through a special link at the bottom lock and most of the lockage water was discharged into them. Only the final 6in was passed into the last stretch of canal and hence to the Oxford. At the same time another stream was tapped to feed into the summit and side ponds were brought into use on the Buckby flight of locks to conserve water[19].

All these works were carried out under Barnes' supervision. Although he had retired he was specially retained to see the Braunston works through to a successful conclusion[20].

In September 1806 land was purchased at Braunston for a new wharf and warehouse near the Daventry road, a short branch canal to serve the wharf being planned around the site of a third reservoir that was needed to provide more capacity for the lockage water. This reservoir was built but the other work was deferred.

The first engine reservoir was small, just over half an acre, and it was linked by an open channel to the second, just over one acre in size. When the third was built it was separated from the others by

Boats on the Braunston summit, at the eastern end of the tunnel, waiting for a stoppage to clear in June 1905. *National Waterways Museum*

In 1898 a new pumping station was built at Braunston near the bottom lock to pump water from the reservoirs back up to the summit, replacing an earlier installation. This picture was taken soon after the work had been completed. *National Waterways Museum*

a footpath and a culvert was provided. The third is larger at nearly two and a half acres. At the time there was never any connection with the main line, and this enabled the Grand Junction and the Oxford to guard their own water supplies[21].

In 1810 the steam engine at Braunston was replaced by a new Boulton & Watt engine which cost over £3,000 and was designed to pump eighty locks per day. It lasted until 1896 when it was scrapped and a Gwynnes steam pump installed in a new engine house built alongside the canal below the bottom lock. At one time there was a plan to extend the channel into which the engine pumped westwards along the contour to pick up several streams, but this was never carried out.

Several other small streams supply the canal. The Bragborough feeder comes in below the third lock at Braunston, the Welton Wharf feeder enters the canal from Welton Place farm in the north, just east of Welton bridge, and the Norton feeder enters from the west at the bottom of Buckby locks. The latter, with another stream, drains land in Norton. The more northerly of the two supplies the canal, whilst the other passes under the canal in a culvert to link up with streams from Buckby and Murcott and eventually to run into the river Nene.

At one time a long channel was planned as an extension of the Watford feeder to enter the canal below Buckby bottom lock almost opposite the Norton feeder. This was to take a roundabout course, running parallel with a stream by way of Murcott, Long Buckby and Whilton mills. Barnes and Barker tried to purchase the mills in April 1805 and divert their water directly into the summit, but they failed, and no action was taken on the proposed feeder either.

In December 1806 the Warwick & Napton Canal was planning a reservoir to hold 14,000 locks at Ashby St Ledgers, whose water would have been fed initially into the Braunston summit. The Oxford opposed the scheme and the Grand Junction imposed so many restrictions that it was dropped in February 1807.

The three-and-a-half-mile-long Braunston summit is considerably deeper than the rest of the canal. In March 1805 Barnes added four courses of brickwork to the two summit locks, put boarding on the gates, and raised the banks and weirs where necessary to enable the level to be raised by 1ft. The summit was originally built with a 2ft freeboard, as against the more usual 1ft for the rest of the canal, and so raising the level presented few problems. This extra depth was to prove invaluable in many dry years that followed.

The opening of the Grand Union Canal in 1814 did not really affect matters. At Watford the new canal climbed over the same ridge of hills that Braunston tunnel was driven through, and this lockage water was passed to the Grand Junction. As most of the new traffic then moved down from the summit, or had to come up to it, there was no advantage to the water supplies. By then the Braunston supplies were on a more secure basis, with the reservoirs, feeders and the pumping station providing adequate capacity.

Opposite page: A gradient profile of the canal showing its water supplies.

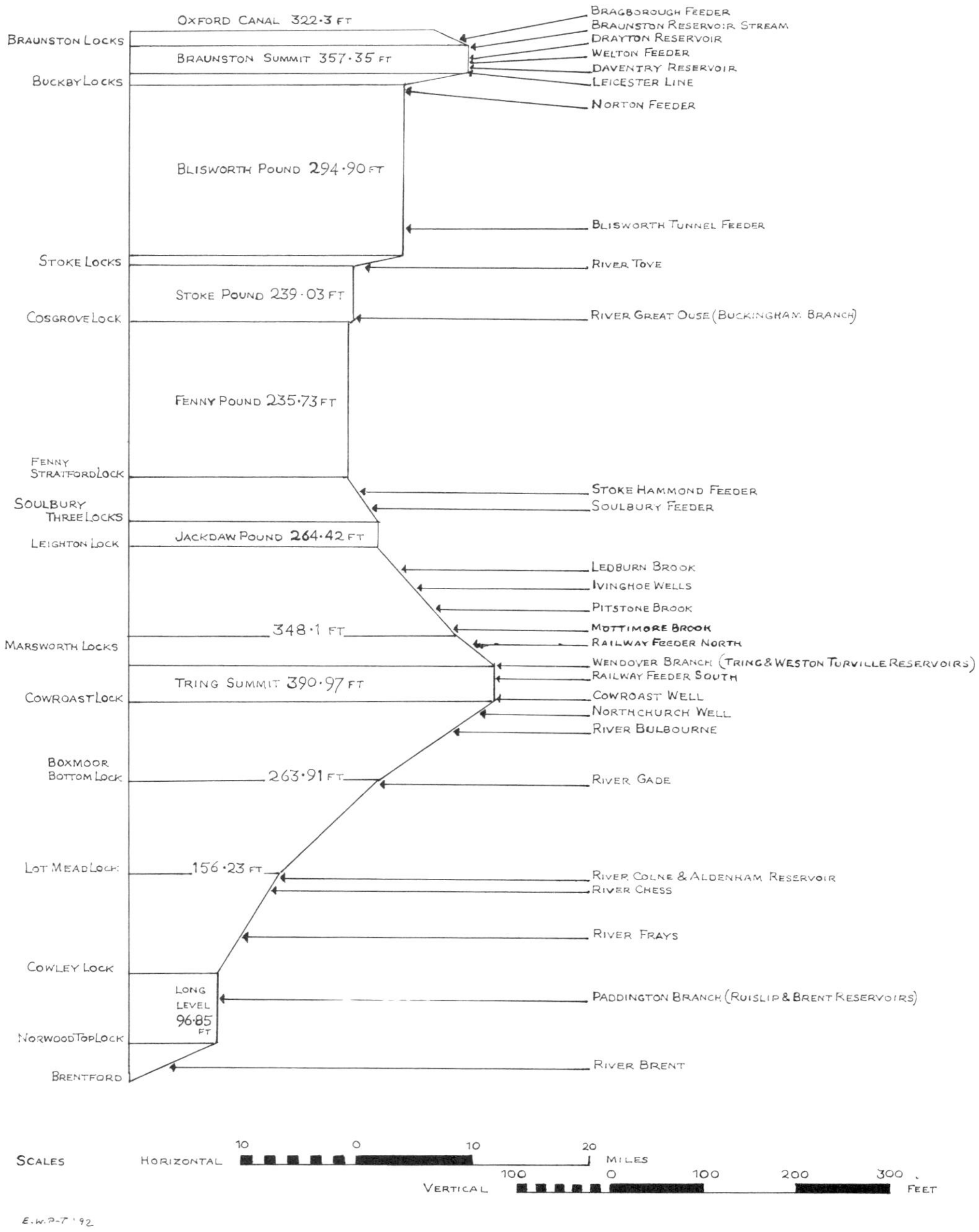
Oxford Canal 322·3 ft
Braunston Locks
Braunston Summit 357·35 ft
Buckby Locks
Blisworth Pound 294·90 ft
Stoke Locks
Stoke Pound 239·03 ft
Cosgrove Lock
Fenny Pound 235·73 ft
Fenny Stratford Lock
Soulbury Three Locks
Leighton Lock
Jackdaw Pound 264·42 ft
348·1 ft
Marsworth Locks
Tring Summit 390·97 ft
Cowroast Lock
Boxmoor Bottom Lock
263·91 ft
Lot Mead Lock
156·23 ft
Cowley Lock
Long Level 96·85 ft
Norwood Top Lock
Brentford
Bragborough Feeder
Braunston Reservoir Stream
Drayton Reservoir
Welton Feeder
Daventry Reservoir
Leicester Line
Norton Feeder
Blisworth Tunnel Feeder
River Tove
River Great Ouse (Buckingham Branch)
Stoke Hammond Feeder
Soulbury Feeder
Ledburn Brook
Ivinghoe Wells
Pitstone Brook
Mottimore Brook
Railway Feeder North
Wendover Branch (Tring & Weston Turville Reservoirs)
Railway Feeder South
Cowroast Well
Northchurch Well
River Bulbourne
River Gade
River Colne & Aldenham Reservoir
River Chess
River Frays
Paddington Branch (Ruislip & Brent Reservoirs)
River Brent
Scales
Horizontal 10 0 10 20 Miles
Vertical 100 0 100 200 300 Feet
E.W.P-T '92

Southwards from Tring

A clause in the company's original Act had specified that a reservoir fed with flood water was to be built to supply water to the river Colne to compensate millers for any diverted into the canal. The company could only take water from the river if it could be made good from the reservoir. To ensure these conditions were kept the millers could appoint a supervisor at the reservoir, although the company had to pay his salary.

A site was selected at Aldenham and the reservoir was one of the earlier works to be started. It had been completed in November 1795, as parts then had to be fenced as they were dangerous to local people. The purchase of the land raised the question of how to divide the money between the lord of the manor at Aldenham and others interested, as it was part of Aldenham common. Just over sixty-eight acres were involved for which the company agreed to pay £30 per acre. Part of the company's sixth Act[22] set out the procedure for settling the matter.

The Act also gave an option to purchase another ten acres at the same price before December 1796. This option was exercised, for in November 1801 Barnes was told to set out the land for the extension. In April 1802 he had to arrange to raise and strengthen the headbank at a cost of £400 for the enlargement.

Aldenham reservoir was fed by several small streams, particularly in the Brockley Hill area to the south. The outlet was down a natural stream into the Colne near Radlett.

The reservoir gave trouble, with slips continually occurring in the headbank. Jessop reported this was due to the treacherous nature of the clay which tended to crack in dry weather, and recommended a 9in-thick protective layer of sand and gravel with a thin covering of soil on the outside slope, sown with rye grass.

In December 1802 Holland had proposed Aldenham should be enlarged again to hold an extra 1,500 locks of water. His estimate was £4,218, and another twenty-one acres would be required. Jessop advised against the plan. He recommended raising the headbank if more water was needed, as this could be done for a third of the cost of the extension. At the time the company was looking into alternative sources of water and no action was taken.

The growth of the Paddington water supply business led to an increased need for water at the southern end of the canal, which demanded an improved supply to the Long Level, the pound running from Cowley to Norwood and Paddington. The most convenient source, the Colne, could not be used and there were restrictions on Aldenham. A fresh source was required.

One scheme was for a new reservoir at Ruislip. Gream had prepared a plan early in 1800, and in April the company started negotiating for the land. Probably for financial reasons little can have been done, and it was not until October 1803 that Barker reopened the discussions.

At the time Ruislip common was being enclosed, and in June 1804 the company applied for land to be allocated for a reservoir. Once this was agreed the work went ahead and the reservoir was brought into service in 1810. The original outlet followed a natural stream and entered the canal not far from Norwood top lock, but this line gave rise to complaints from the Colne millers. In December 1815 to meet these objections it was decided to build a new and shorter feeder to the Paddington branch. Later the final stretch of the old feeder was made navigable and became known as Passmore's Dock.

Even with Ruislip in service supplies to the Long Level were inadequate. The company was also unhappy about the restrictions on its use of Aldenham, whilst the millers were worried about possible abuses with the company taking water from the Colne. The Duke of Northumberland, who owned many of the mills on the Colne downstream from Uxbridge, was particularly concerned to ensure river water was not diverted to supply the Paddington waterworks.

The canal and the river had several junctions north of Uxbridge and, in theory, it would have been a simple matter for the company to feed river water through Cowley lock to the Long Level. The company was specifically debarred from doing this by its first Act, but to safeguard the Duke's interests, and those of his miller tenants, another Act[23] was passed in 1812, prohibiting any new connections being made with the Colne after June 1811. The millers were empowered to appoint a supervisor at both Cowley and Cowroast

Cowley lock was particularly important to the Grand Junction as it controlled the supply of water to the Long Level. Here the motor barge *Uxbridge*, owned by British Portland Cement Manufacturers Ltd, leaves the lock heading northwards, probably bound for the company's works at Harefield. *National Waterways Museum*

locks to control the water passing to the Long Level. Records had to be kept of all craft to determine whether the lockage water at Cowley was passed to the Long Level or to the Colne through a special new link with the river at Cowley.

If a boat was passing through Cowley to, or from, anywhere short of the Tring summit at Cowroast its lockage water could not be supplied from the summit but would have been drawn from the Colne or one of its tributaries. This lockage water had to be returned to the river at Cowley to make good the abstraction. If the boat either went as far as the summit, or had come from it, this water could be passed to the Long Level. The company was also entitled to lockage water for boats passing through Cowroast to, or from, any point north of Cowley.

The Act also amended the restrictions on Aldenham. It laid down that the company had to supply the Colne with 320 locks of water from the reservoir between June and October each year as a soakage and evaporation allowance. Gauges were provided at various places to calculate the amounts of water passing these points. Any other water in the reservoir could be used to supply the Long Level. Whilst these operations had to be carried out under the strict supervision of the millers' representatives at Aldenham and Cowley,

The Norwood and Hanwell locks had to be supplied with water from the Long Level. Here a Maypole Margarine Works barge has just left one of the Hanwell locks and is passing the entrance to Hanwell Asylum Dock in 1926. *National Waterways Museum*

A section of the Long Level at Southall with a horsedrawn pair of narrow boats and other craft unloading at the wharves. *Railway & Canal Historical Society*

they were a great improvement for the company and meant Aldenham could now be used to the full.

For many years there were two lockkeepers at Cowley and Cowroast, both paid by the company. One acted for the company, the other for the Duke of Northumberland. In time, and with the decline in the number of mills and their reliance on water power, the duties of the Duke's men were assumed by the company's keepers, with the Cowroast man reporting the appropriate numbers to his Cowley colleague.

Another possible source of water was from the river Brent. In June 1803 Barnes suggested a site for a reservoir and Jessop was asked to comment. He supported the idea and Barker then drew up a plan and estimate of the land needed, but in October 1803 the project was deferred on cost grounds.

So pressing was the water situation that in April 1804 Jessop was asked to report on a proposed feeder from the river at Kingsbury through Willesden to the Paddington branch at Lower Place. Still no action was taken, but the scheme was revived in 1809. Henry Provis carried out a new survey which differed little from the 1804 plan. The feeder was to be three and a quarter miles long.

The company now decided to go ahead and opened negotiations for the land required. The works were not heavy and the feeder was soon in service. Later it was incorporated into the Brent reservoir scheme which was carried out by the Regent's Canal Company and opened in June 1835. Most of the feeder was used for the outlet from the reservoir.

As well as finding new sources of water the company was careful to conserve existing supplies. In December 1803 two experimental side ponds were built at locks 58 and 59 just south of Berkhamsted. Bevan reported in April 1805 on their costs of construction and the extra time needed to work through these locks. It was then decided to install side ponds at key locations to conserve water.

An obvious place was at Norwood and Hanwell, where locks carried the canal down from the Long Level to the river Brent. Any saving here would relieve pressure on the supplies to Paddington. Construction was deferred, but in 1815 John Woodhouse contracted to build the eight side ponds needed. In December Provis reported his work was so bad that the ponds were unusable. Woodhouse carried out repairs and eventually the problem was sorted out.

With the reservoirs and water supplies now reasonably secure the company and its officers were better placed to meet the next major challenge.

NOTES TO CHAPTER SIX

1 All the Grand Junction water tables were measured in locks. A lock is the equivalent of 56,000 gallons.
2 Richardson, Alan. Water Supplies to Tring Summit. *Journal of the Railway & Canal Historical Society,* April 1969.
3 Ibid.
4 McGarey, D.G. *The History of Tringford Pumping Station,* 1946.
5 52 Geo III, c 140—9 June 1812.
6 Water Supplies to Tring Summit, op cit.
7 *History of Tringford Pumping Station,* op cit.
8 Ibid.
9 Ibid.
10 The capacity of Wilstone reservoirs was reduced when the overflow weir was dropped by 12in in November 1936.
11 Thomas, Gordon C. Evidence to Royal Commission on Canals and Inland Waterways, 1909. (Reply 41346)
12 Ibid, 41351.
13 Ibid, 41351.
14 Water Supplies to Tring Summit, op cit.
15 Evans, Joan. *The Endless Web,* 1955.
16 Water Supplies to Tring Summit, op cit.
17 Report to General Assembly, 7 June 1803.
18 The overflow weir has now been lowered by 23in, making the present capacity 5,500 locks.
19 *Northampton Mercury,* 31 August 1805.
20 Barnes died on 18 January 1819, aged 79. There is a monument to him in Bodicote church.
21 Since nationalisation these reservoirs have been connected to the canal.
22 36 Geo III, c 25—24 December 1795.
23 52 Geo III, c 140—9 June 1812.

The Railway Challenge 7

THE DEVELOPMENT of a network of railways throughout the country presented a major threat to the Grand Junction and to every other canal. As early as January 1823 the company had been approached by the Leeds & Liverpool Canal Company seeking support in opposing a plan for a railway from Manchester to Liverpool. At the time railways were in their infancy, but the successful opening of the Stockton & Darlington line in September 1825, and the subsequent proving of the merits of the steam locomotive, heralded an enormous challenge to the canals. Whilst the waterways system had represented a major improvement in the existing transport facilities of the day, it was not ideal for the increasing role it was being called upon to play.

One of the main drawbacks was that the system had developed piecemeal with a multitude of companies and no overall co-ordinating policy. Most of the Midlands canals had been built to a standard size capable of taking a 7ft-wide narrow boat. The Grand Junction had been built as a 14ft canal but the narrow locks on the routes to Coventry and Birmingham prevented the through working of barges. Similarly the narrow locks on the Grand Union Canal prevented barges reaching Leicester, whilst those on the Northampton branch prevented barges reaching the river Nene.

There were sound financial reasons for building to these gauges at the time, but faced with competition from newer forms of transport, they were considerable handicaps. The canal companies, lacking any common policy, were incapable of ridding themselves of these problems. There was also no real effort to simplify the toll system. Long-distance traffic was faced with a bewildering number of different companies, each with its own toll structure, which made quoting through rates difficult.

Relations between different companies were often poor. Throughout their history, the Grand Junction and the Oxford looked on one another with mutual distrust, and most attempts at co-operation foundered. Whilst the canals enjoyed a monopoly the damage caused by this disunity was not apparent, but in a competitive situation it was very different.

The first railway plan to affect the Grand Junction directly was in 1824[1], and involved a line from London to Birmingham. It proved premature, but a few years later the plan was revived and by the summer of 1830 two projects were being actively canvassed[2].

The first, surveyed by Francis Giles, was for a line through Coventry, Rugby and Hemel Hempstead to terminate at Islington. The second, surveyed by Sir John Rennie, passed through Banbury and Oxford. George and Robert Stephenson were called in to investigate both routes and they recommended the Rugby line, whereupon the two groups joined forces to promote an Act, with Robert Stephenson as the engineer. His route differed little from Giles' line, and as such it presented a major threat to the Grand Junction.

At the time an average of twenty-six fly boats daily in each direction carried the lighter and more perishable traffics[3]. These were likely to be the first to transfer to the railway, which promised to be so much faster. Meanwhile there were far greater numbers of ordinary boats carrying the bulk cargoes.

When the London & Birmingham Bill was presented to the Commons at the end of 1831 it met with a storm of vigorous opposition. This was led by the Grand Junction, as the railway was planned to run parallel with it for many miles. Many landowners were active opponents, and as the busy Watling Street was also directly threatened, road coach proprietors, waggoners and their supporting industries such as posting inns and stables, all joined in the outcry.

At the beginning of 1832 the Grand Junction attempted to unite action from all the canal companies involved. Even in the hour of danger there was disunity, with the Birmingham Canal Company refusing to subscribe to the general opposition expenses. The company, finding the landowners were its most powerful allies, allocated £2,000 towards a joint fighting fund.

Much depended on the outcome, for if the Bill succeeded it would pave the way for further trunk lines throughout the country. Despite the opposition the Commons passed the Bill in June 1832, but the Lords threw it out in July[4]. This provided only a temporary respite, for the Bill was re-presented in the next session. In the meantime the railway company was able to buy off most of the objecting landowners[5]. As a result the opposition crumbled and the Act for the London & Birmingham Railway (L&BR) was passed on 6 May 1833.

It was a great tribute to Jessop and Barnes that the L&BR followed the canal line so closely. Like the canal it was to climb to a massive cutting through the Chilterns at Tring. There then followed a long descent to Wolverton, where the Great Ouse was crossed by a large embankment and viaduct. Perhaps forewarned by the Grand Junction's struggle at Blisworth, Stephenson decided to drive a cutting through the same ridge at nearby Roade. This proved a wise decision, but it still took a long and arduous struggle to complete the work.

It was further north at Kilsby, where Stephenson intended to drive a tunnel through the same hills pierced by Braunston tunnel and by Crick tunnel on the Grand Union, that major problems arose. It seems he knew about the difficulties encountered at Crick (see Chapter 8) and moved his line south to avoid the quicksands found there. He seems not to have heard about the same quicksands at Braunston, and the struggle that ensued at Kilsby soon matched the Blisworth saga.

Opposite page: The butty *Daphne,* belonging to Aylesbury carrier Arthur Harvey-Taylor, passing through one of the sixteen locks on the six-and-a-quarter-mile Aylesbury branch, which joins the main line at Marsworth. Best known of the Aylesbury carriers, Harvey-Taylor took over an existing company, whose black and white livery he retained, in 1923. He acquired the *Daphne* in 1937 from Joseph Grantham. Harvey-Taylor's boats, which carried coal, grain, sand, aluminium and strawboard, were in 1955 absorbed into the Samuel Barlow fleet.

DAPHNE

Despite the difficulties, the line from the London terminus at Euston to Boxmoor was opened in July 1837; from Boxmoor to Tring in October; from Tring to Bletchley in April 1838; and finally through Roade cutting and Kilsby tunnel to Rugby in September. The line from Birmingham to Rugby was finished in April 1838.

The railway crossed the canal six times: at Long Buckby, Blisworth, Wolverton, Cheddington, Winkwell and Nash Mill. In several places the line ran particularly close to the canal: Long Buckby to Weedon; Bugbrooke to Blisworth; Soulbury to Leighton Buzzard; and a long stretch from the Tring summit down to Hunton Bridge.

There was a major confrontation at Wolverton in 1834. The L&BR wanted to build a temporary bridge, which involved driving piles into the canal banks. The Grand Junction disputed the right to erect this bridge. Despite this Stephenson assembled a large workforce and began building the bridge on the night of 23 December. Work continued non-stop until it was finished by noon on Christmas day. The Grand Junction was not prepared to ignore this provocative challenge. On 30 December its Northern district engineer, Thomas Lake, led a strong body of canal employees in demolishing the new bridge[6].

The dispute went to the Court of Chancery in January, when the L&BR obtained an injunction to prevent the Grand Junction destroying any of its works. Following an appeal by the L&BR in February 1835, an extension from fifteen to twenty-eight days was agreed in the time allowed to build bridges over the canal.

At Weedon, where the L&BR crossed the military dock, an unusual rolling bridge was provided, mounted on its own rails and manhandled diagonally out of the way when a boat needed to pass, leaving a gaping hole in the railway[7]. Apart from the delays to canal traffic the situation was far from ideal for the railway. In 1888 the London & North Western Railway (L&NWR), as the L&BR had become in 1846, diverted the line westwards on to a new bridge which provided just sufficient headroom for boats to pass. At the same time a siding was built into the depot, but gunpowder and live ammunition continued to be handled by canal until after the First World War.

Once the L&BR was built other lines followed. A branch from Cheddington to Aylesbury, running virtually parallel with the Aylesbury branch canal, was opened in June 1839. Six years later a branch was opened from Blisworth to Northampton, parallel with the Northampton branch canal. In November 1846 a line was opened from Bletchley to Bedford, crossing the canal at Fenny Stratford, and in May 1850 another opened from Bletchley, through Buckingham, to Banbury. Further south a line from Leighton Buzzard to Dunstable opened in May 1848.

In July 1840 the Midland Counties Railway between Leicester and the L&BR at Rugby was finished, providing a direct link between the Derbyshire and Nottinghamshire coalfield and

Opposite page: A northbound train headed by one of the London & North Western Railway "Claughton" class express locomotives crosses the iron arched skew bridge north of Kings Langley on the eve of the outbreak of war in 1914, seen from a pair of John Dickinson's "paper dashers" heading for Nash mills with a cargo of pulp. The London & Birmingham Railway, which in 1846 became the London & North Western, crossed the Grand Junction six times.

Opposite page: The course of the London & Birmingham Railway complemented that of the Grand Junction Canal. This map is based on the position in 1853.

London. Another link was forged in April 1857, when the Midland Railway from Bedford to the Great Northern Railway at Hitchin was opened. The carriage of coal to both Rugby and Hitchin was soon very heavy; so much so that the Midland opened its own direct line from Bedford to London in 1868.

From the Midland at Harpenden a line opened to Hemel Hempstead in 1877. It crossed the canal just below Boxmoor bottom lock and can only have taken traffic, mainly in coal, from the Grand Junction. Further south a stretch of the Great Western Railway (GWR) had been opened in June 1838 almost parallel to the canal between Paddington and Yiewsley. A branch from Yiewsley to Uxbridge opened in September 1856.

The Brentford Railway from Southall to Brentford followed in July 1859. It included a dock off the river Thames to transfer goods between barges and the railway. When the line opened the Grand Junction calculated that it would lose £4,200 at the Brentford and Bulls Bridge toll stations. The line was noteworthy in that it passed under the canal at the same place as the canal passed under Windmill Lane in Norwood; the canal had to be diverted whilst an aqueduct was built there.

The GWR built a dock on the canal at Hayes near the present Hayes & Harlington station; from 1856 regular transhipment began from the railway to barges for destinations along the Paddington branch and on the Regent's Canal. The GWR also established a sleeper depot alongside the Paddington branch at Southall, supplied

The Great Western Railway maintained a sleeper depot beside the Paddington branch at Southall. Most of the timber was received by barge from the London Docks, and at one time most of the creosote came by narrow boat from the Midlands. *National Waterways Museum*

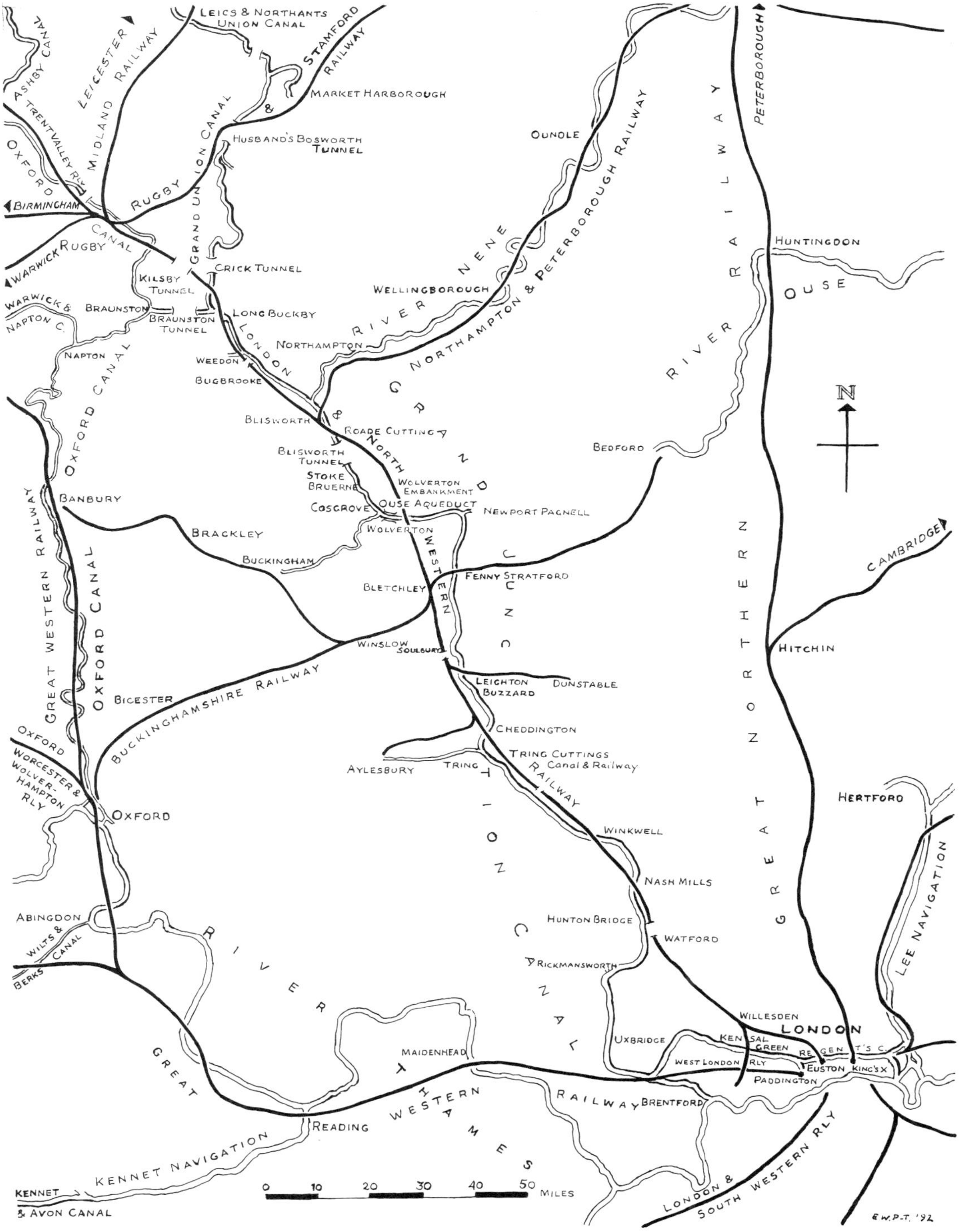

LEICS & NORTHANTS UNION CANAL
STAMFORD & RAILWAY
MARKET HARBOROUGH
LEICESTER
MIDLAND RAILWAY
ASHBY CANAL
TRENT VALLEY RLY
OXFORD
BIRMINGHAM
RUGBY
CANAL
WARWICK
GRAND UNION CANAL
HUSBAND'S BOSWORTH TUNNEL
CRICK TUNNEL
KILSBY TUNNEL
WARWICK & NAPTON C.
BRAUNSTON
BRAUNSTON TUNNEL
LONG BUCKBY
NAPTON
OXFORD CANAL
WEEDON
BUGBROOKE
NORTHAMPTON
LONDON & NORTH WESTERN
RIVER NENE
WELLINGBOROUGH
OUNDLE
NORTHAMPTON & PETERBOROUGH RAILWAY
PETERBOROUGH
RAILWAY
HUNTINGDON
OUSE
RIVER
N
BLISWORTH
ROADE CUTTING
BLISWORTH TUNNEL
STOKE BRUERNE
WOLVERTON EMBANKMENT
COSGROVE
OUSE AQUEDUCT
NEWPORT PAGNELL
WOLVERTON
BEDFORD
BANBURY
BRACKLEY
BUCKINGHAM
GRAND JUNCTION CANAL
FENNY STRATFORD
BLETCHLEY
WINSLOW
SOULBURY
GREAT WESTERN RAILWAY
OXFORD CANAL
BICESTER
BUCKINGHAMSHIRE RAILWAY
LEIGHTON BUZZARD
DUNSTABLE
CHEDDINGTON
TRING CUTTINGS Canal & Railway
AYLESBURY
TRING
RAILWAY
GREAT NORTHERN
CAMBRIDGE
HITCHIN
OXFORD WORCESTER & WOLVERHAMPTON RLY
OXFORD
HERTFORD
WINKWELL
NASH MILLS
ABINGDON
WILTS & BERKS CANAL
RIVER
HUNTON BRIDGE
WATFORD
RICKMANSWORTH
LEE NAVIGATION
WILLESDEN
LONDON
UXBRIDGE
KENSAL GREEN
REGENT'S C.
MAIDENHEAD
WEST LONDON RLY
EUSTON
KING'S X
PADDINGTON
GREAT WESTERN RAILWAY
BRENTFORD
THAMES
READING
KENNET NAVIGATION
KENNET & AVON CANAL
0 10 20 30 40 50 MILES
LONDON & SOUTH WESTERN RLY
E.W.P.-T. '92

with timber brought up the canal by barge from the London docks, and with creosote brought by narrow boat from the Midlands.

The railway's first gains were in the passenger and perishable goods traffic. Passenger movement had never been significant on the Grand Junction, but goods requiring a speedy transit formed an important part of its trade. The railway had an obvious advantage here. For the heavier bulk traffics speed was less important, and the two rivals were soon involved in a rates battle.

The company's tonnage revenue had peaked in 1836. From 1 January 1837 substantial toll reductions were brought in, causing the revenue to drop by nearly a quarter. General goods were reduced by 25% from 84½p to 63p per ton, and coal dropped to 0.3p (three farthings) per ton mile, but with a maximum of 22p per ton, or 10½p if brought all the way to London. At the same time the regulations were eased, with the canal now open twenty-four hours a day and mixed cargoes being permitted. Further reductions became necessary later.

The drop in revenue would have been even more severe had it not been for an upsurge in traffic. This arose from the ever-growing demand for transport from increasing industrialisation, and particularly in traffic from the Black Country centred on Birmingham to London. The railways captured much of this growth, but the Grand Junction retained its position as an important artery of trade.

A steam tug emerges from the eastern end of Braunston tunnel with a train of boats in tow whilst the horses, which had been walked over the top of the tunnel, wait patiently on the towpath. *National Waterways Museum*

An example of the growth of industrial traffic was in iron from Staffordshire and Shropshire carried by canal to London[8].

	1815	*1820*	*1825*	*1830*	*1835*	*1840*
Manufacturing iron	14,900	17,722	15,258	17,338	25,427	43,613
Pig iron	2,832	5,208	3,471	1,289	420	1,254
Cast iron pipes	2,475	445	905	–	–	–
Castings	3,240	8,237	9,901	8,205	13,542	10,736
Wrought iron bedsteads	–	–	581	58	4	91
Totals	23,447	31,612	30,116	26,890	39,393	55,694

Coal traffic from Braunston to London was 72,256 tons in 1844, but despite toll cuts this fell to 60,311 in 1845[9]. In that year the railway handled only a minute quantity—a mere 8,377 tons due to the initial reluctance of the L&BR to carry it at all. This reluctance was soon overcome, whereupon the railway's tonnage increased sharply—to 55,096 in 1850 (against the canal's 29,479), to 1,137,835 in 1855 (23,251), and to 1,477,547 in 1860 (19,593). By 1867 the tonnage brought by rail, 3,295,652, mainly by the L&NWR and the Great Northern Railway, exceeded the sea coal tonnage, which had been fairly static between two and three million tons, for the first time[10]. By now the canal's share had fallen below 10,000 tons.

Accurate figures were kept because of the Coal Duties Acts, to which the railways were also subject. The boundary for these duties was originally computed as being twenty-five miles from the General Post Office in London. In 1845 the distance was reduced to twenty miles, but the Grand Junction collector remained stationed at Grove. In 1861 the area was amended again to the Metropolitan police district. On the canal this meant the boundary was now a short way below Stockers lock, and a stone obelisk was erected beside the towpath to mark the position.

The collector then moved from Grove to a new house built for him beside the lock[11]. By now the money raised on canal-borne coal was falling rapidly in line with the tonnage. In August 1863 it yielded £36 against £7,132 for rail-borne and £14,552 for sea-borne coal. At the end of 1869 the appointment of the Stockers collector was discontinued, with the company undertaking to collect the dues for an annual fee of £35[12]. The coal dues finally came to an end in 1890.

A coal duties marker beside the towpath.

One effect of the railways was to stimulate interest in an improved canal from London to Liverpool. Part of the plan, the Birmingham & Liverpool Junction Canal, was opened in 1835, and in November 1827 plans for a London & Birmingham Junction Canal were announced[13]. It was to run from the Stratford-upon-Avon Canal, about ten miles south of Birmingham, to the Grand Junction at Braunston. It was to pass under the Oxford, and there was to be a branch to the Coventry. Telford was the engineer.

The main advantage of the plan was that it avoided the heavy

The London & Birmingham Railway originally used stone block sleepers for its track. When these were replaced by timber sleepers the Grand Junction purchased several thousand of the blocks and used them for protecting the canal banks, as shown here. *Author's collection*

lockage into, and out of, the Avon valley at Warwick; only twenty locks were needed against nearly eighty by the Warwick canals. The Coventry branch would have reduced the through route to the north by twenty miles, bypassing the circuitous northern section of the Oxford. The plan was strongly opposed by the Oxford, the two Warwick canals, and the Trent & Mersey, who all feared loss of traffic.

The Oxford, however, was prompted into straightening and improving its northern section and, as a result, the London & Birmingham Junction was amended to run to the Oxford at Ansty[14]. In February 1830, with the terminus now shown as Brinklow, a Bill was introduced, but it was thrown out in May. The plan did not die, although Telford ceased to be involved. His place was taken by William Cubitt, who had been involved with the works on the Oxford. In 1832 he produced plans for a Central Union Canal to run from the Worcester & Birmingham Canal in Birmingham to Bordesley on the Warwick & Birmingham, and then from that canal at Solihull to the Oxford at Ansty. The plans got nowhere.

At the end of 1833 a group of promoters, backed by Lord Dudley and other Staffordshire coal and iron masters who opposed the restrictive toll agreements in force between the Coventry and the Warwick canals, planned the London & Birmingham Canal. The line, surveyed by James Green, was to run from the Stratford-upon-Avon Canal through Banbury, Buckingham, Tring, and St Albans to Highgate, where a branch would link up with the Regent's Canal. It was to be 113 miles long with only forty-eight locks, but with a further twenty-two on the Regent's branch.

From the Stratford there were to be nineteen locks and then a level summit pound of over seventy miles to St Albans. Nearly ten miles of tunnelling would be needed, together with an enormous aqueduct over the Avon valley. The cost was estimated at £3,000,000[15]. A prospectus, issued in March 1836, envisaged an elaborate scheme with twin tunnels, each with a towing path. The sides of the canal were to be walled and a double towing path provided throughout. It was claimed goods would be delivered in London in thirty-two hours, instead of seventy by the existing route.

The Grand Junction opposed the plan, and in 1836 compromise proposals were made involving junctions at either Stoke Bruerne, Weedon or Braunston, and using the Grand Junction line on to London. These were all rejected, as was a plan to join the Buckingham branch in the following year[16]. Other canal companies also opposed the scheme. As it was the promoters were unable to raise the capital needed, as the current vogue was investment in railways. There was a final meeting in Cubitt's London office in March 1838 but thereafter the plan was shelved.

Just as the earlier London & Birmingham Junction scheme had spurred the Oxford to improve part of its line, so this latest project stimulated the Grand Junction into making improvements.

One was to double the Stoke Bruerne flight of locks to speed up the traffic and to save water in the long Blisworth pound, which had

The bridge immediately below Stoke Bruerne top lock has two arches, as at one time there were duplicate locks here to cope with the volume of traffic. The downstream approach to the duplicate lock has been blocked off, but the bridge survives and is now a listed building. *Alan Faulkner*

to supply the Northampton branch as well. Detailed plans were drawn up by William Thompson and John Holland and work started in March 1835. Duplicate wide locks were provided along the west side of each of the seven existing locks, whilst the road bridges at the tail of the top lock in Stoke Bruerne village and below lock 18 were rebuilt with a double waterway arch.

In January 1838 land was purchased at Marsworth to build narrow locks in parallel to locks 37 to 45 on the northern descent from the Tring summit. William Cubitt tendered for the work in June, but his figure was too high. Grissell & Peto were then asked to tender, and they got the job.

After the Marsworth locks had been dealt with the company decided to extend the duplication all the way down to Stoke Hammond; the whole series was complete by the middle of 1839. Six bridges, all either immediately above or below a lock, had to be provided with a double waterway arch, and most of these remain to this day. In all twenty-three locks northwards from Tring were duplicated, but Fenny Stratford was omitted, probably because of its very small rise.

The main reason for this duplication was to speed up single narrow boats, which made up a high proportion of the craft using the canal at the time. In the original Act boats less than 60ft in length, 12ft in width and carrying under thirty tons, could be prevented from using the canal unless there was a surplus of water. When there was a water shortage single narrow boats often had to wait until they could find a partner to pass through the wide locks, unless they were prepared to pay extra charges. The number of single boats was substantial and in December 1836 a new reduced tariff was introduced which relaxed the restrictions. In the years that followed whenever there was a water shortage, as in December 1837, the restrictions were reimposed, but they did not apply where the narrow locks could be used.

In October 1837 the company also considered providing narrow locks southwards from the Tring summit to Hunton Bridge, but disagreements with the millers at Kings Langley prevented the scheme being carried out.

None of the duplicate locks had a long life. The water supply situation at Tring soon improved, once the third Wilstone reservoir and the chain of pumps from Fenny Stratford were in service. There was also a trend away from single boats to pairs. After a while the locks north of Tring were filled in, and with modern bank protection work most traces of them have disappeared. Similarly at Stoke five of the duplicates were filled in, but at the top and bottom of the flight they were converted into side ponds. New side ponds were provided for the other Stoke locks; the working drawings were prepared in 1852 by Jesse Cherry, the Northern district engineer.

Mounting railway competition soon highlighted the lack of unity among the various canal companies and the poor relations between the Grand Junction and the Oxford. At the end of the 1830s the Grand Junction was seeking co-operation from other companies with toll cuts to stimulate traffic between London and Manchester. At first it had little success, the others blaming the acute water shortages on the Tring summit for the difficulties the trade was then facing.

In May 1841, when the Grand Junction's water position had improved, there was a meeting of all the companies except the

This water tower beside the railway at Blisworth, pictured in May 1957, was supplied from the canal. Blisworth is just one of six places where the London & Birmingham Railway crossed the canal.
Alan Faulkner

Oxford, which refused to attend. The plan was for a standard through toll of 0.1p (one farthing) per ton mile for the entire 249-mile journey. At the time the Bridgewater, Trent & Mersey, Coventry and Grand Junction were all charging double this. The Birmingham, which controlled a five-and-a-half-mile section at Fradley, charged four times, and the Oxford, with a twenty-three-and-a-half-mile section, six times the proposed rate.

Without the co-operation of the Oxford, and to a lesser extent the Birmingham, no progress was made. The minutes of the meeting were sent to the Oxford, but to no effect. At another meeting, to which the Oxford at last sent delegates, the Grand Junction claimed the Oxford's ton mileage charge on coal was seven times its own. The Oxford countered that the Grand Junction's was a special rate for the London trade only, that its own rate was based on the original mileage, and that no extra charge had ever been made for the convenience of using its shortened line.

The Oxford also felt the Grand Junction favoured the Leicester canals through its special relationship with the Grand Union. The two Leicester companies were usually willing to co-operate with the Grand Junction—from 1 May 1845 a general reduction to 0.2p (one halfpenny) per ton mile was agreed on all goods to and from the Leicester line. This placed goods entering at Braunston at a disadvantage, as the Grand Junction refused to extend the new rates to them until the Oxford agreed to reduce theirs.

The Grand Junction kept trying, and in May 1846 approached the Warwick Canals with a view to a merger. They were willing to negotiate and a new canal to bypass the Oxford was considered. Thomas Lake estimated £72,914 would be needed for a canal just over six miles long from the third lock at Braunston, with an aqueduct over the Oxford, to join the Warwick & Napton (W&N) near Napton.

These plans disturbed the Oxford, as did a suggestion for a similar merger with the Leicester canals, and a new flat rate of 0.1p (one farthing) per ton mile all the way to London, but with all traffic leaving the Oxford being charged at the same rate as the Oxford was then charging. Whilst this would have harmed the Grand Junction, it would have crippled the Oxford by stopping the thriving coal trade from the Warwickshire collieries.

After more procrastinating the Oxford relented and a comprehensive toll agreement was reached following a series of arduous negotiations between Sir Francis Head, the Grand Junction's chairman, and Dr David Durrell, the Oxford's chairman.

Meanwhile the downward trend in tolls continued and an attempt to increase the Birmingham toll to 63p per ton in May 1849 resulted in Shiptons, one of the larger carriers, leaving the canal. They were enticed back at the end of the year with a 50% reduction, but it was a short-lived return as Shiptons carried their last freight to and from London in the middle of 1853. Their tolls had produced £3,014 in 1852[17].

In November 1851 the W&N compensation toll was abolished. This dated back to the very early days, when the W&N, or the Warwick & Braunston as it was originally called, was to have joined the Grand Junction at Braunston, thus completely by-passing the Oxford on the London to Birmingham route.

In 1795 the Warwick & Braunston applied to Parliament to alter its line to join the Oxford at Napton, saving four miles of new canal and £62,000 in expense. The Grand Junction applied for its £10,000 guarantee, awarded to the Oxford in its original Act, to be modified, but it offered to drop the request if the Warwick & Braunston was built as a wide canal, as more revenue would accrue[18]. As this could not be agreed the guarantee was left unchanged, but the Grand Junction was awarded a 2½p compensation toll on goods passing between the W&N and its own line.

To collect this toll the Grand Junction opened a station on the W&N at Napton near the junction with the Oxford. The £10,000 guarantee was repealed in the 1829 Act authorising the shortening of the Oxford's line, and came to an end on 31 March 1832.

The railways had a marked effect on the price of the Grand Junction's shares. In its earliest days these had a speculative value, and often attracted a large premium; in September 1792 ten shares of £100 each, with only £1 paid up, were sold for £450[19]. The early financial problems caused the price to drop, and by 1800 the shares were being exchanged at face value. There was a small rise over the next few years and by 1807 a modest premium of £10 per share was being attracted.

Thereafter the demand, and hence the price, moved steadily upwards in line with the increasing dividend; to £240 in 1810, and £320 in 1811. There was a drop in 1816 when the dividend was cut, but by 1824 a record price of almost £350 was being obtained. The price declined to £312 in 1827 and to £295 in 1829, with a sharp fall to £236 in 1830 as investors turned to railway schemes. Ten years later the price was half this, but thereafter it rallied and remained around £100 for many years.

Similarly the dividend maintained its peak of 13% for seven

The diversion of the canal at Northampton carried out for the London & North Western Railway Company.

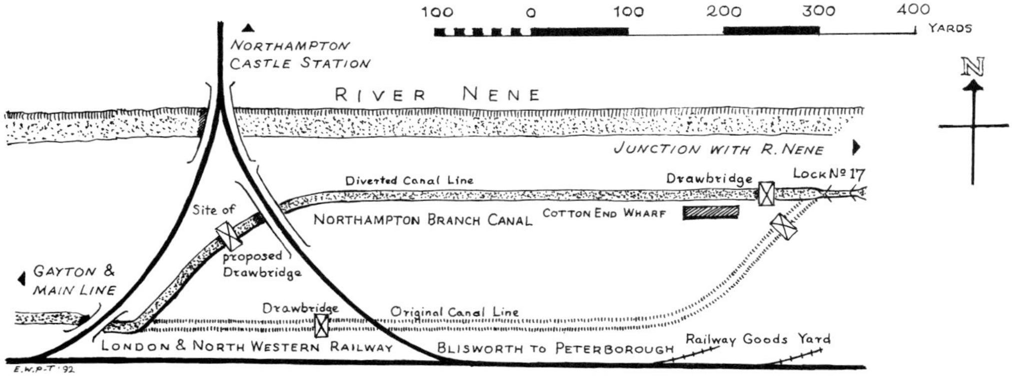

years until 1838, when it dropped to 10%. It then tumbled to 5% in 1847 and to 2% in 1855, but it stabilised back at 4% in the 1860s.

This drop reflected the heavy rate cutting by both the canal and the railways, but in 1857 an agreement was reached with both the L&NWR and the GWR, the two most concerned with the Grand Junction, to end the rate-cutting and stabilise tolls. On the Grand Junction most goods between Paddington and Braunston were now charged at 0.1p (one farthing) per ton mile, or 10½p for the full 101-mile journey.

Two small sections of the Grand Junction were taken for conversion to a railway. Under an agreement with the Birmingham, Bristol & Thames Junction Railway, later the West London Railway, a portion of the Paddington branch at Wormwood Scrubbs was used, whilst the railway purchased other land on which to divert the canal. The work was duly carried out, but the railway was dilatory in completing the purchase of the land for the canal. It had still not paid in January 1842, when the original owners served an ejectment notice on the Grand Junction. This notice was withdrawn after a few weeks, with the aggrieved owners seeking redress from the guilty railway company. At first the West London line passed underneath the canal, but in 1858 the line was diverted northwards to cross over it instead.

The other section was at Northampton. In 1876 the L&NWR obtained powers to construct a connecting line to enable trains from its main Northampton Castle station to link up with the

The FMC boat *Orange* sunk nearly opposite the ironstone loading stage just north of Blisworth tunnel. The ironstone was brought in wagons from quarries east of the canal, across the bridge here, to await transhipment.
National Waterways Museum

Blisworth to Peterborough line in an easterly direction, which involved diverting a 1,050yd stretch of the Northampton branch. The new canal was brought into service in June 1879, the railway paying all the costs and £1,500 as compensation.

There was a different involvement with the railways at Blisworth. Ironstone had been discovered in the area and the first shipments were sent out by canal to the Midlands in 1852. Soon afterwards a quarry opened near the north end of Blisworth tunnel and a tramway was built from this, crossing over the canal on a bridge, to a loading wharf. The ironstone was then taken by water the short distance to a new siding at Blisworth station for transhipment to the L&NWR[20]. At least two other quarries sent out ironstone to the railway in the same way. Later much of the ironstone was boated down the Northampton branch to be unloaded at Hunsbury Hill Furnaces instead, and this lasted until the furnaces closed in 1921. Hunsbury Hill lock was appropriately known as Furnace lock.

Another way in which the Grand Junction tried to compete with the railways was in carrying. At first most canal companies were reluctant to carry on their own waterways since by so doing it was possible for them to give preferential rates. That would have been equivalent to varying the tolls, which was then illegal. Although the

One of the ironstone gantries at Blisworth with an open boat fitted out with small containers being loaded. The ironstone was taken either to Blisworth station for transhipment to the railway or down the Northampton branch to Hunsbury Hill Furnaces. *George Freeston*

Grand Junction had always owned trading boats, these were hired out to other traders. In 1845 the legal position was amended by the Canal Carriers Acts[21], and at the same time varying the tolls was made legal.

At first the Grand Junction was disinclined to change its policy. In June 1847 Thomas Grahame, a strong advocate of carrying, took over as chairman, and a new select committee was elected, almost certainly due to internal disagreements as to whether the company should set up as carriers. Grahame's appointment was timely as Pickfords, who were still the most important firm on the canal,

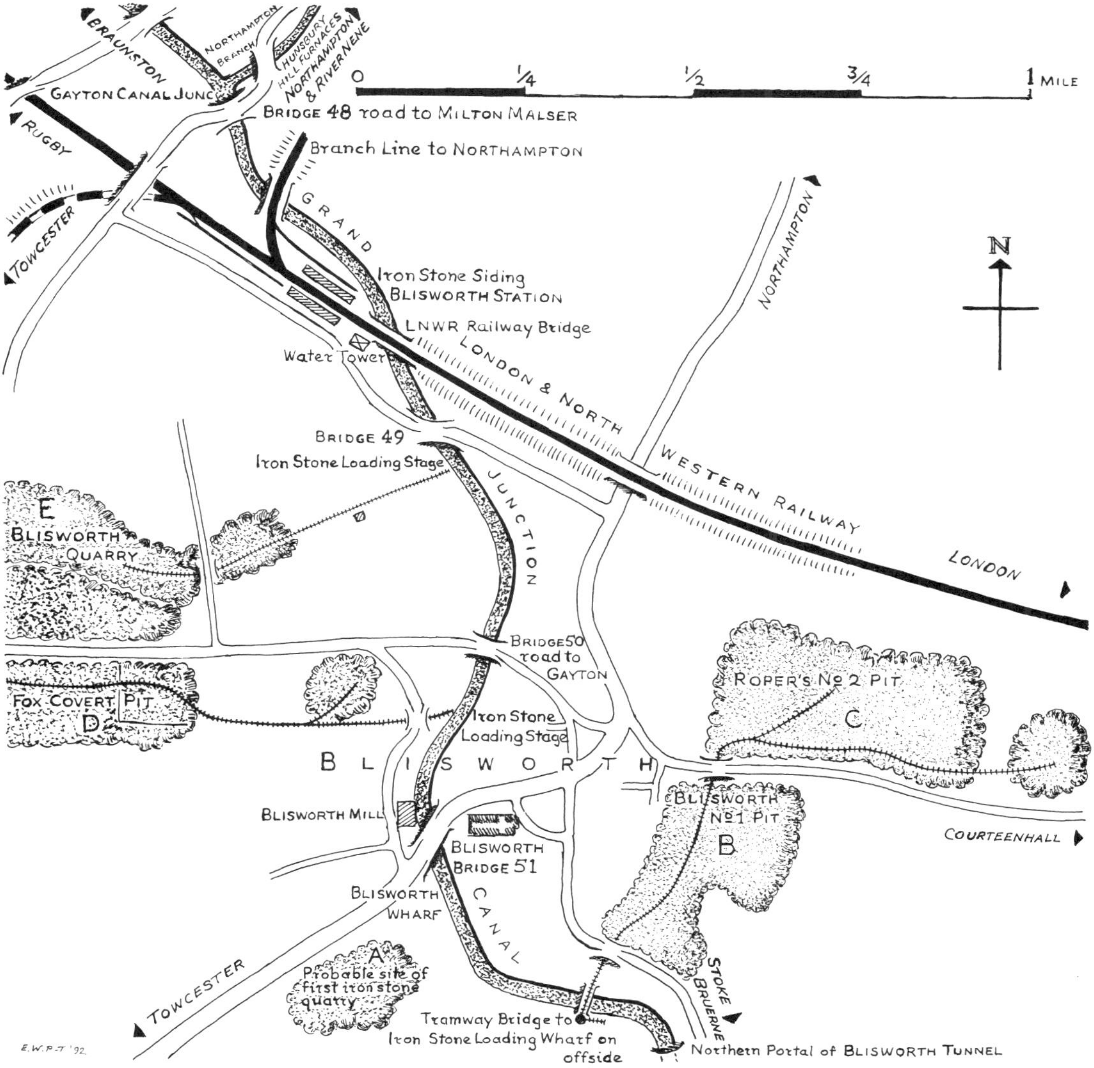

The canal and the Blisworth ironstone quarries.

Blisworth mill was an important user of canal transport. It was near to this mill that the leggers waited for boats to be taken through the tunnel. *National Waterways Museum*

decided to give up canal carrying at the end of 1847 and go over to the railways.

The Grand Junction had already appointed a manager for the canal and the proposed carrying department in August, and the potential loss of such a large part of its trade removed any final objections. The shareholders now agreed to form a Carrying Establishment. Special toll rates were secured from the companies on the Warwick and Leicester lines, and negotiations were opened to take over most of Pickfords' boats and traffics and several other carriers as well.

To finance the acquisitions £114,550 of a new 6% preference stock was issued to the existing shareholders in the proportion of £10 of new stock for each £100 ordinary share held. An initial call of 25% was made on 1 September 1848, and the other three soon afterwards.

A significant proportion of the Grand Junction's total trade was soon being handled by the Carrying Establishment. In 1851 it contributed £29,100 to a total revenue of £79,600, and this rose to a peak of £31,700 in 1853. Despite this, carrying was not profitable and the manager left in the autumn of 1850, following a £12,800 carrying loss for the first half of the year. A proposal to reduce the dividend to 1% was defeated on the questionable grounds that the carrying side was now on a firmer basis.

Although consideration had been given to reducing the size of the carrying business, it remained substantial. In June 1853 the company had 510 boat horses and 136 cart horses. It handled 85,750 tons in the half year to 25 June and earned £80,500, compared with 68,260 tons and £68,300 in the same period of 1852.

Whilst economies were made the Carrying Establishment still

Leggers emerging from Blisworth tunnel. *National Waterways Museum*

made a loss, and in 1864 a proposed merger with the Warwick canals was called off because they were unwilling for the enlarged concern to continue carrying. The Grand Junction was reluctant to give up because it had been successful in maintaining traffic, and much of this trade could be lost.

In 1835 total tonnage had been 824,674, of which through traffic was 192,859. Thereafter through traffic increased to a peak of 294,257 tons in 1845, but had declined to 135,657 by 1870. Overall traffic showed a steady increase to 1,031,284 tons in 1848, 1,142,450 in 1858, and 1,404,012 in 1868.

Following a disastrous explosion in 1874 on the Regent's Canal involving some of the company's boats, and as losses were still being incurred, carrying was given up in 1876. All that remained were the preference shares. For some time the company had been buying these in for cancellation; by June 1863 they had reduced to £100,000. Ultimately they reduced to £93,700, at which level they remained as a legacy of the ill-fated carrying venture.

An unusual problem arose at Rickmansworth in 1836. William Dimes, a City solicitor, had acquired the Lordship of the Manor of Rickmansworth in May 1831. It included the freehold reversion of some copyhold land purchased by the company from the tenant, Joseph Skidmore, when the canal had been built. Copyhold land was held by a tenant during his lifetime, but on his death it could revert to the Lord of the Manor. The company had never had the transfer registered in the manorial records, and had never been admitted as the new tenant[22].

Skidmore died in 1835 and, as no relative claimed the land, Dimes started an action in 1836 to eject the company. The action failed but Dimes appealed. This appeal was heard on 7 June 1838

Opposite page: Close to Bulls Bridge junction where the line from Paddington meets the main route from Brentford was the Great Western Railway's sleeper depot, alongside the Paddington branch and served by boats bringing timber. Passing by is the wide boat *Swallow,* one of the St. Marylebone Borough Council fleet which carried domestic rubbish from central London to tips on the outskirts; she was built in 1923 and incidentally docked by Walkers' of Rickmansworth in 1948 for extensive repairs.

and was successful. Dimes then erected a barrier across the towpath and obstructed the canal to prevent boats passing.

The company obtained a temporary injunction against him on 18 June and re-opened the canal. Dimes claimed his interest in the land was worth £10,000, although he would accept £5,000 if it was paid immediately. Against this the company offered £3,560, and on paying half this sum into court the injunction was extended indefinitely.

Meanwhile the company's solicitors had traced Skidmore's son and heir, Thomas. As he would have been liable, under an indemnity given by his father, to refund the original purchase price, he agreed to apply for the tenancy of the disputed land as nominee for the Grand Junction. The company agreed to pay all the costs and any fines, which together might total £1,500.

Dimes refused to accept Thomas's claim and the dispute went back to court. In November 1846 judgement was given that Dimes would have to admit Thomas as the new tenant. An appeal followed and in April 1849 Dimes discovered the judge involved had a small shareholding in the Grand Junction, which meant he could not give an impartial ruling. Dimes contended this made the judgement void and in January 1850 he attempted to close the canal again. The company responded by having him imprisoned for infringing the 1838 injunction. More litigation followed and Dimes appealed to the House of Lords, which found for the company in July 1852.

In November the company agreed to purchase the freehold for £700, and in April 1853 it received the deed of enfranchisement, giving it the right to the uninterrupted use of its canal, seventeen years after the problem arose.

Another problem was Blisworth tunnel, which needed constant repair. In 1849 a length of brickwork was found badly distorted, and a series of iron ribs, or struts, were installed in October until permanent repairs could be carried out. They remained in place until 1854, but thereafter there were few years when major works of some kind were not needed. In 1864 two special cross dams were designed and built to permit repair work on sections of the invert to be carried out without having to drain the entire tunnel.

When the Braunston and Blisworth tunnels were opened all traffic was horsedrawn and the boats had to be shafted or legged through, whilst the horses were walked over the top. Legging consisted of men lying at the ends of two planks, fixed across the fore-end of the boat, and pushing with their feet against the tunnel walls. In 1827 gangs of professional leggers were employed by the company. They were issued with a brass armlet and were known as "registered leggers". Before that the system was on a casual basis and boatmen were often terrorised to force them to employ leggers.

At Blisworth the men were paid 7½p per trip. They only handled traffic one way, the Blisworth men working southbound and the Stoke men northbound, with four men from each gang on duty at night[23]. Huts were provided at each end of the tunnel where the men waited for the boats, one next to the Boat Inn at Stoke, the

St MARYLEBONE
No 31
BOROUGH COUNCIL
GJC 12109
LOCO
G W

DRINK
DELICIOUS
OVALTI
FOR
HEALTH

other opposite Blisworth mill. The bridge nearby at Blisworth was called Candle bridge, for the boatmen used to buy tallow dips there to light them through the tunnel.

A steam tug boat, similar to the one used at Islington tunnel on the Regent's Canal, had been considered for both Blisworth and Braunston in July 1844, but nothing was done. In July 1870 a wire rope haulage system was suggested for both tunnels, and the company's engineer was authorised to install it, initially only at Braunston on trial. The work was completed in October, and a house for the steam engine to drive the system was built at the eastern end of the tunnel. The wire rope was carried along the wall of the tunnel on pulleys, round a large pulley at the western end, and then back down the other side. A 5p charge was made for its use.

By the end of November the system had been judged a failure, due to excessive friction caused by the double bend in the tunnel and the great strain on the wire. Early in 1872 the company accepted a £50 offer for some of the pulleys and wire rope that had been used.

In February 1871 steam tugs were provided at Blisworth, and two months later they were laid on at Braunston. The tugs worked to a regular timetable, with a charge of 7½p for boats carrying twenty-five tons or over, 6p for those with less than twenty-five tons, and 5p for empty boats. At Blisworth seventeen leggers were paid off, and five others found work elsewhere on the canal. At Braunston the company paid a 25p per week pension to three of the leggers including one man, then seventy-five years old, who had been at the tunnel for forty-four years, and was one of the original leggers from when the registered system had first started.

Steam-driven craft soon gave rise to ventilation problems, particularly in the long Blisworth tunnel. This was highlighted by an accident in September 1861, when two men suffocated whilst a steamer was passing through the tunnel[24]. Within a week one of the old construction shafts in the middle of the tunnel had been reopened and a request made to the Duke of Grafton to open two more. There were more complaints in 1864, and again in June 1881, when another two shafts were opened. Seven now act as ventilators[25].

Another problem caused by steam-driven craft was the layer of soot that built up on the tunnel roof, becoming so thick its removal was essential. At first the soot was cleared by using a hawthorn bush, tied to the top of a boat that was dragged through, but later a special brushing boat was devised. Three wire brushes, shaped to the profile of the tunnel, were attached to long arms and mounted on a trestle structure on a dredging boat. The boat was towed through the tunnel, the brushes being kept in contact with the roof by men pressing down on the opposite ends of the handles, and the soot dropping down into the boat's hopper. The tunnel was then swept out regularly by this strange boat.

Throughout the early years of the railway age the company's

Opposite page: Fumes and smoke drift from the southern portal of Blisworth tunnel, as they did continuously in days of commercial traffic, providing a contrast to the purity of the Ovaltine advertisement on the cratch of a butty belonging to the fleet of A. Wander Ltd, manufacturers of the well-known night-time drink. Walkers' at Rickmansworth built eight motors and eight butties for Wander between 1927 and 1929.

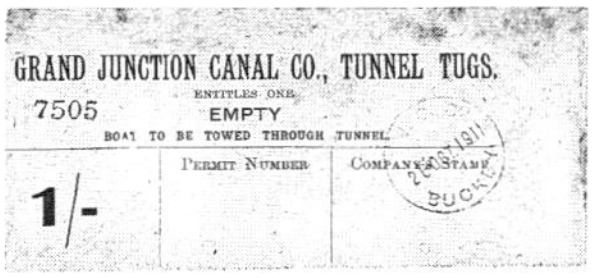

A Braunston tunnel tug ticket dated 26 October 1911. *National Waterways Museum*

chairman was Philip Pleydell Bouverie, and for most of his term in office he acted as general manager, latterly receiving a £750 salary. He was succeeded by Sir Francis Bond Head at the beginning of June 1840, who also acted as general manager, initially at the same salary, but later this was doubled. In 1847 he was replaced by Thomas Grahame, who in turn was succeeded by George Anderson in 1854. Anderson remained in the chair until 1868, when the Hon Robert Howe Brown took over. He served until his death in 1888.

The office of clerk was filled by Richard Sale until his death in June 1845, after a lifetime of service with the company. He was replaced by Thomas Chrimes, who had to resign after four months owing to ill health. To fill the gap Charles Rogers, appointed the company's accountant in December 1842 when George Jackson retired, became acting clerk. His appointment was made permanent in 1847.

In the early days the clerk was concerned mainly with legal matters, but as time went on more and more of the day-to-day management of the canal fell on his shoulders. Rogers retired in 1869 and was elected to the select committee. His place was taken by Charles Mercer, who died in October 1886 and was succeeded by his son Charles. Charles had to resign in 1891 following an investigation into deficiencies in the accounts.

On the engineering side John Holland was dismissed as Southern district engineer in December 1838, because he had been persuading suppliers to give false invoices and receipts for payments not made, and then claiming reimbursement from the company. He was replaced by John Lake, but in January 1841 the district was extended northwards to Berkhamsted, and William Brown was

The Blisworth brushing boat about to enter the northern end of the tunnel. The brushes were designed to fit the crown of the arch and were kept in contact with the brickwork by the men pressing down on the handles. *National Waterways Museum*

appointed. John Lake transferred to the Middle district, which now covered from Berkhamsted to Fenny Stratford. It had had no permanent engineer since 1835, when Thomas Lake, John's father, had transferred to the Northern district. In July 1839 the district had been temporarily divided, with William Haynes taking charge of the section from Hunton Bridge to Marsworth, and Thomas Lake covering from Marsworth to Leighton Buzzard, in addition to the Northern district.

The January 1841 alterations also affected the Northern district, which was divided into two sections: from Fenny Stratford to the bottom of Buckby locks with Haynes in charge; and from Buckby to Braunston with Thomas Lake in charge. As senior engineer, Thomas Lake also acted as consultant for the whole length of the canal.

William Brown was dismissed in July 1843, and soon afterwards John Lake took back the Southern district, whilst retaining the Middle district. In the north Jesse Cherry took over a reunited Northern district from Thomas Lake and William Haynes in January 1851. Cherry died in 1875 having served fifty-eight years

The steam tug *Anslow* at the southern end of Blisworth tunnel in the early 1930s. It was named after Tonman Mosley, a director of the Grand Junction from 1894 to 1928, who was created first Baron Anslow in 1916. *National Waterways Museum*

with the company. John Lake died in 1864, and the company then appointed Hubert Thomas as engineer for the entire canal. His starting salary was £300, but this was soon increased and had reached £1,000 by 1885. He served until 1891, when he became clerk to the company, his son Gordon becoming engineer. It was during their reign that there were important developments on the Grand Junction.

NOTES TO CHAPTER SEVEN

1 Jeaffreson, J.C. & Pole, W. *The Life of Robert Stephenson,* 1864.
2 LMS Railway Company. *Old Euston,* 1938.
3 Rolt, L.T.C. *George & Robert Stephenson,* 1960.
4 *Old Euston,* op cit.
5 *George & Robert Stephenson,* op cit.
6 Ibid.
7 Compton, Hugh & Faulkner, Alan. Weedon Military Dock. *Railway & Canal Historical Society Journal,* November 1991.
8 Scrivener, Harry. *A Comprehensive History of the Iron Trade,* 1841.
9 Grand Junction Toll Receipt Books, 1841–75.
10 *An Epitome of the Progress of the Trade in Coal to London since 1775.* (1869 edition).
11 Bawtree, Maurice. The City of London Coal Duties Boundary Marks, in *The London Archaeologist,* spring 1969.
12 Bawtree, Maurice. Stockers House and the London Coal Duties Boundary Marks, in *Rickmansworth Historian,* autumn 1965.
13 *Aris's Birmingham Gazette,* 12 November 1827.
14 Hadfield, Charles. *Canals of the East Midlands,* 1966.
15 London to Birmingham Canal, in *Lock & Quay,* December 1954.
16 *Canals of the East Midlands,* op cit.
17 Grand Junction Toll Receipt Books, 1841–75.
18 *Case of the Grand Junction Company respecting the Variation of the Warwick & Braunston Canal* (printed leaflet c. February 1796).
19 *Northampton Mercury,* 1 September 1792.
20 Tonks, Eric. *The Ironstone Quarries of the Midlands, Part III, The Northampton Area,* 1989.
21 Canals Carriers Acts (8 & 9 Vic, c 28 and c 42; 10 & 11 Vic, c 94).
22 Sharman, Frank. Dimes v The Grand Junction Canal Company. *Railway & Canal Historical Society Journal,* July 1986.
23 Rolt, L.T.C. *The Inland Waterways of England,* 1950.
24 *Rugby Advertiser,* 14 September 1861.
25 From the northern end of Blisworth tunnel there are ventilators at 441, 982, 1249, 1746, 2035, 2174 and 2722yd. The deepest shaft is 164ft from the top to water level. At Braunston from the western end there are ventilators at 278, 373, 762, 1208 and 1481yd, but two are now closed. The deepest shaft is 156ft from the top to water level.

The Leicester Line 8

The river Trent has always been a major artery for trade in the East Midlands. For many years boats could go no farther inland than Nottingham, but after the Civil War the navigation was extended to Burton, and by the early 1720s its tributary, the Derwent, had been made navigable to Derby[1].

Similar attempts at the time to make the Soar, another major tributary, navigable from the Trent at Redhill southwards, through Loughborough to Leicester, failed. They were revived later and by the middle of 1777 the Loughborough Navigation[2], a nine-mile canalisation of the river, was opened. The undertaking was soon highly profitable.

At the same time there was a complementary scheme for a canal to run from the Trent near Redhill northwards up the Erewash valley to Langley Mill. Collieries in the area could then send their coal down the Erewash Canal, across the Trent and up the Soar to Loughborough, from where road transport would be used to supply Leicester. The Erewash Canal[3] was opened in 1779 and also became successful.

In 1791 an Act[4] was obtained to extend the Soar navigation nearly sixteen miles southwards to Leicester. The Leicester Navigation was opened formally in October 1794 and a busy trade soon developed, mainly in coal from the Derbyshire and Nottinghamshire collieries, supplemented from the west Leicestershire coalfield. This was linked to the navigation at Loughborough by the part-canal, part-railroad, Charnwood Forest line.

With trade growing on the Soar, suggestions were soon made to extend the navigation still further southwards. Market Harborough was an obvious target, and early in 1792 a meeting was held in the town[5] to consider making a canal from Leicester. The scheme had influential backing and attracted much general support.

By the middle of 1792 a more ambitious plan was being considered to extend the canal beyond Market Harborough to link up with the river Nene and the Grand Junction at Northampton[6]. A survey was made by Christopher Staveley and initially the two extensions—from Leicester to Market Harborough, and from Market Harborough to Northampton—were in the hands of separate committees with separate subscription lists.

In the summer of 1792 William Jessop was appointed engineer, with John Varley to carry out detailed surveys. Varley at first recommended linking with the Grand Junction at Gayton, and a branch from there to the Nene. The Grand Junction, however, was promoting its own branch from Gayton to the Nene and the proposed

Opposite page: The Leicester Line, showing its connecting waterways.

link was altered to a point on its branch near Northampton.

A further extension had been suggested from Northampton to link up with the river Lee at Hertford, and connecting with the Great Ouse at Bedford[7]. This idea was abandoned in favour of using the Grand Junction's line for the through route to London.

There was little opposition to the plan, and an Act[8] for the Leicestershire and Northamptonshire Union Canal (L&NUC), or the Old Union Canal as it was more popularly referred to later, was obtained on the same day as the Grand Junction's Act. It authorised making the Soar navigable from West Bridge in Leicester upstream to Aylestone, and from there a canal to the Grand Junction at Hardingstone, with a link to the Nene near Northampton.

The main line was to be almost forty-four miles long, with a branch of nearly four miles from Lubenham to Market Harborough. Apart from a series of locks out of the Soar valley and down into the Nene valley the main works were four tunnels: Saddington 880yd, Foxton 1,056yd, Kelmarsh 990yd, and Great Oxendon 286yd. The Act authorised raising £200,000 in £100 shares, with a further £100,000 if needed.

Work started from Leicester with Varley and Staveley in charge. On 28 October 1794 the first five miles were opened to Blaby through eight locks[9]. By March 1795 a further three miles through four locks was finished to Kilby. Progress then slowed down as the company was finding it difficult to raise funds fast enough, and was also having to contend with labour disputes. In October 1795 it decided to do no work south of Gumley.

Contracts for Saddington tunnel were let in July 1795. The work was completed in February 1797, some delay having been caused when it was found the tunnel was not perfectly straight. Three short sections had to be widened slightly to allow barges to pass through safely. On 7 April 1797 the canal was opened to Debdale wharf near Gumley[10], seventeen miles from Leicester through twenty-five locks. Work then stopped, although a reservoir was completed at Saddington in 1800 to supply the canal with water.

Saddington tunnel on the Old Union Canal between Leicester and Market Harborough. There is a tunnel-top path for the horses.

In 1799 James Barnes was called in to advise on linking the unfinished canal to the Grand Junction at Braunston. Nothing was done, but in 1802 he made another survey for a line through Welford and Crick to Norton on the Grand Junction's Braunston summit. His estimate was almost £147,000, which included a branch to Market Harborough.

Telford was called in to advise on Barnes' plan, and he recommended a much more roundabout route, with fewer locks, using part of the original line through Market Harborough, but still ending up at Norton. His estimate was £228,000, which omitted a proposed branch to Rushton, about three miles north of Kettering[11].

Still nothing was done until 1804, when the company decided to vary the line between Gumley and Market Harborough to avoid the long tunnel at Foxton. A second Act[12] was obtained in 1805.

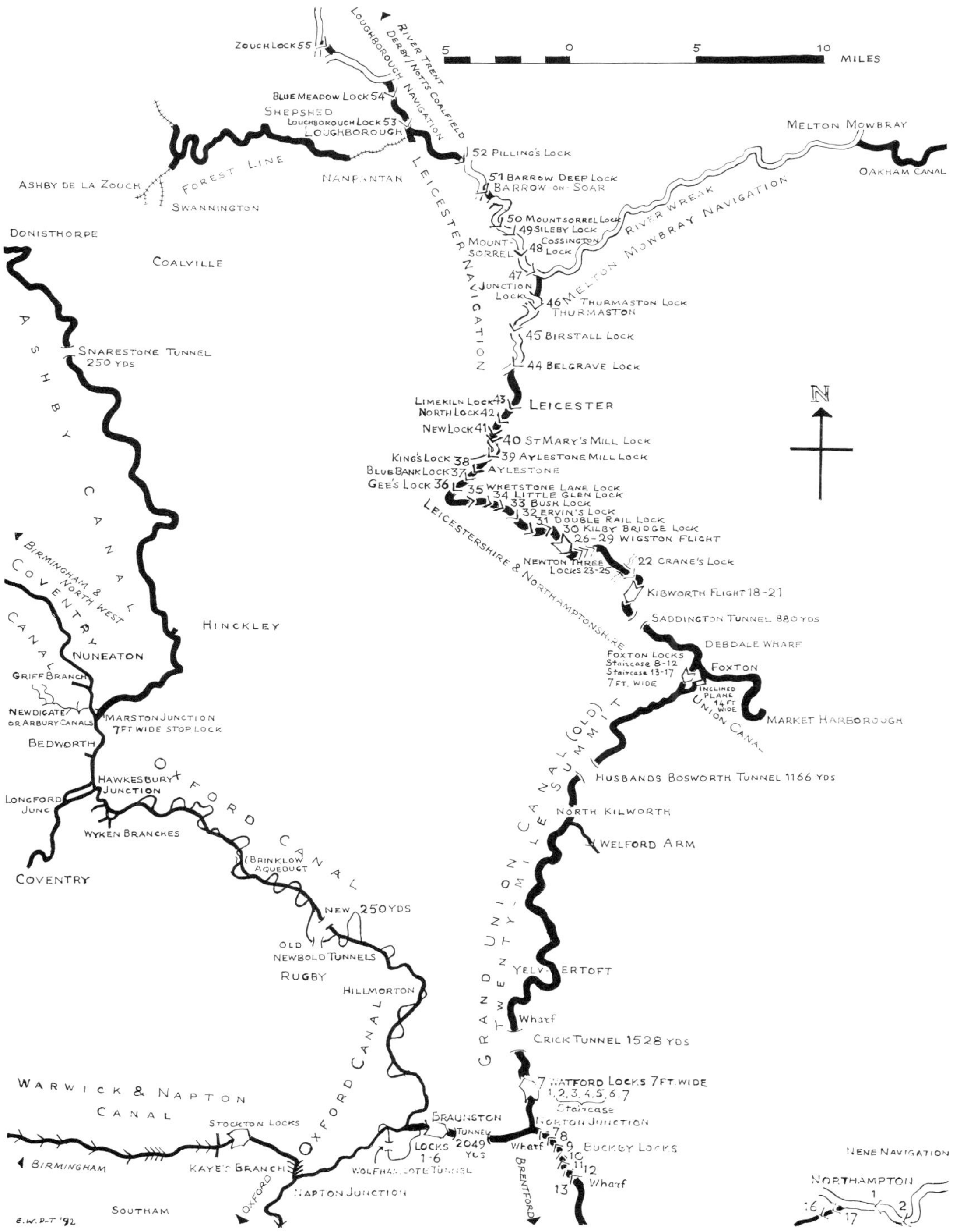
5 0 5 10 MILES
N
ZOUCH LOCK 55
RIVER TRENT
DERBY / NOTTS COALFIELD
LOUGHBOROUGH NAVIGATION
BLUE MEADOW LOCK 54
SHEPSHED
LOUGHBOROUGH LOCK 53
LOUGHBOROUGH
52 PILLING'S LOCK
MELTON MOWBRAY
OAKHAM CANAL
ASHBY DE LA ZOUCH
FOREST LINE
NANPANTAN
SWANNINGTON
LEICESTER NAVIGATION
51 BARROW DEEP LOCK
BARROW-ON-SOAR
50 MOUNTSORREL LOCK
49 SILEBY LOCK
COSSINGTON
48 LOCK
MOUNT-SORREL
RIVER WREAK
MELTON MOWBRAY NAVIGATION
DONISTHORPE
COALVILLE
47
JUNCTION LOCK
46 THURMASTON LOCK
THURMASTON
45 BIRSTALL LOCK
44 BELGRAVE LOCK
SNARESTONE TUNNEL 250 YDS
ASHBY CANAL
LIMEKILN LOCK 43
NORTH LOCK 42
LEICESTER
NEW LOCK 41
40 ST MARY'S MILL LOCK
39 AYLESTONE MILL LOCK
KING'S LOCK 38
BLUE BANK LOCK 37
AYLESTONE
GEE'S LOCK 36
35 WHETSTONE LANE LOCK
34 LITTLE GLEN LOCK
33 BUSH LOCK
32 ERVIN'S LOCK
31 DOUBLE RAIL LOCK
30 KILBY BRIDGE LOCK
26-29 WIGSTON FLIGHT
NEWTON THREE LOCKS 23-25
22 CRANE'S LOCK
KIBWORTH FLIGHT 18-21
LEICESTERSHIRE & NORTHAMPTONSHIRE
SADDINGTON TUNNEL 880 YDS
DEBDALE WHARF
BIRMINGHAM & NORTH WEST
COVENTRY CANAL
HINCKLEY
NUNEATON
GRIFF BRANCH
NEWDIGATE OR ARBURY CANALS
MARSTON JUNCTION
7FT WIDE STOP LOCK
BEDWORTH
FOXTON LOCKS
Staircase 8-12
Staircase 13-17
7FT. WIDE
FOXTON
INCLINED PLANE 14 FT WIDE
UNION CANAL
MARKET HARBOROUGH
(OLD) UNION CANAL
SUMMIT
HUSBANDS BOSWORTH TUNNEL 1166 YDS
HAWKESBURY JUNCTION
OXFORD CANAL
LONGFORD JUNC
WYKEN BRANCHES
NORTH KILWORTH
WELFORD ARM
COVENTRY
(BRINKLOW AQUEDUCT
NEW 250 YDS
OLD
NEWBOLD TUNNELS
RUGBY
TWENTY MILE
GRAND UNION CANAL
YELVERTOFT
HILLMORTON
Wharf
CRICK TUNNEL 1528 YDS
WARWICK & NAPTON CANAL
OXFORD CANAL
7 WATFORD LOCKS 7FT. WIDE
1, 2, 3, 4, 5, 6, 7
Staircase
BRAUNSTON
NORTON JUNCTION
STOCKTON LOCKS
TUNNEL 2049 YDS
LOCKS 1-6
Wharf
8 BUCKBY LOCKS
NENE NAVIGATION
BIRMINGHAM
KAYE'S BRANCH
WOLFHAMCOTE TUNNEL
BRENTFORD
10
11
12
Wharf
13
NORTHAMPTON
1
2
16
17
NAPTON JUNCTION
OXFORD
SOUTHAM
E.W.P-T '92

Even then, it was not until 13 October 1809 that the six-and-three-quarter-mile level extension to Market Harborough was opened with considerable ceremony[13]. It had cost over £40,000. In 1807 the company had investigated the possibility of continuing to the Grand Junction, but in May 1808 the decision was taken to terminate at Market Harborough[14].

Following a meeting in London in June 1808 the Grand Junction had encouraged a separate group of promoters to explore making such a link. The L&NUC was content to support this new initiative. Benjamin Bevan was called in as surveyor, and his first task was to examine the previous plans and select the best route. Apart from the lines suggested by Telford and Barnes, there was another for a canal from Blaby to Cosford, near Rugby, on the Oxford.

By the end of 1808 Bevan favoured following Barnes' line from Foxton through Welford to Crick, but then recommended avoiding the tunnel by linking with the Oxford near Braunston[15]. It is certain this did not suit the Grand Junction at all, and in May 1809 a line was adopted with a tunnel at Crick and then running to Norton. It was virtually the same route Barnes had put forward in 1802. The estimate for the Grand Union, as the new canal was to be called, was now £205,000.

The Grand Junction was enthusiastic about the plan. Not only would it connect with the Soar, the Trent, and other waterways, but it could also be a part of a shorter canal route to Manchester by using the Erewash, the Cromford, and a proposed canal across the Peak to join the Peak Forest Canal.

This latter objective was strongly opposed by the Bridgewater and Trent & Mersey Canal Companies, who retaliated by petitioning against the Grand Union Bill. They only withdrew when the Grand Junction withdrew its support for the Peak canal project.

Other opponents were the Oxford, which was eventually indemnified against loss of trade; Northampton and the river Nene Commissioners, who forced concessions for the Northampton branch canal; and Aylesbury, which also obtained an agreement for its branch canal. The Grand Union Act[16] was passed in May 1810. It authorised raising £245,000 in shares, with a further £50,000 if needed.

Leggers emerging from the north-eastern portal of Husbands Bosworth tunnel, opened in 1813.

Construction of the new canal began from the north at Foxton, where there was a flight of ten locks, in two staircases of five, lifting the canal up to its twenty-mile-long summit level. By June 1812 five miles of canal were open from the top of these locks, where a temporary railroad connected with the L&NUC, to the cutting at the north end of Husbands Bosworth tunnel[17]. On 1 October Foxton locks were finished, and part of the line south from Husbands Bosworth was also open. Eight shafts were used during the construction of the 1,170yd tunnel. The brickwork was completed on 29 April 1813, and on 25 May the deep cutting to the south was completed and the canal opened to Stanford, about ten miles from

Foxton, along with the first mile of the Welford branch to Welford mill[18].

Meanwhile Bevan had found quicksands and other poor strata on the line of the proposed 1,000yd tunnel at Crick, and in 1812 the original line was abandoned. A revised route was adopted running to the east of Crick village, which involved a longer 1,528yd tunnel, but overall the canal would be 700yd shorter.

There was originally to have been a third tunnel, some 400yd long, with a deep cutting at Watford above the locks. In October 1813 the company managed to avoid this by paying £2,000 compensation to divert the line through Watford park. At Watford the descent from the summit level was made by a group of seven locks, of which four were in a staircase.

The Grand Union Canal (GUC) was opened on 9 August 1814[19]. It was just over twenty-three miles long, with seventeen locks. Early in 1815 Welford lock was completed, and the short branch of one and five-eighth miles opened to Welford wharf. This also served as a feeder to supply water from reservoirs on Naseby Wolds.

The GUC locks were built narrow, thus breaking the chain of wide locks between London, the East Midlands and the North East. At the time the Grand Junction was opposed to barges and wide

A scene at Welford Wharf about 1907. Mrs Mary Ann Gilbert was the main carrier to Welford; she is seen here with her two daughters, one of whom was named Gwen Mary. *Alan Squires*

Foxton locks were grouped in two five-rise staircases with a passing bay between the two. They provide a dramatic climb of 75ft., with the top lock cottage silhouetted on the skyline.

boats passing through its long Blisworth tunnel, as this would have involved a system of one-way traffic and the delays would be severe. At the same time most of the carriers likely to use the new through route had forecast they would use narrow boats.

The decision to build narrow locks produced a major saving in cost and water consumption, but it was a decision that gave rise to much controversy later. The bridges and tunnels were built to a 14ft gauge so that barges could have used them if the locks had permitted.

The L&NUC paid a maiden dividend in 1812 and made regular payments thereafter. They rose to a peak of 6% in 1837, but then declined, with a particularly sharp drop in 1851. By 1855 the dividend was down to 2.125% and to 0.5% by 1871.

The GUC was in a much weaker position for, unlike the L&NUC which had some local business, it depended almost entirely on through trade for its revenue, and was entirely at the mercy of other companies on the route. It also had a large debt burden which was not finally paid off until 1836. A first dividend of 1% was paid in 1827, but this did not rise until 1840, and then only to 1.75%. A decline then set in with the dividend down to 0.5% by 1865, and to a tenth of this in 1885.

From 1863 the GUC entered into an informal arrangement with the L&NUC whereby the two were worked effectively as a single company. Throughout their existence both concerns enjoyed good relations with the Grand Junction, which at one time had a major shareholding in the GUC. Both were usually ready to co-operate over toll adjustments without the frustrating negotiations that characterised approaches to the Oxford. In 1865 the Grand Junction had a disagreement with the GUC over the tolls paid by its Carrying Establishment, but this was far from typical.

In 1881 Leicester Corporation obtained an Improvement Act[20] enabling it to widen and straighten the Soar through the city to relieve flooding. The first two L&NUC locks, at Castle mill and Swan's mill, were demolished and replaced by a new lock in 1888. The entire project was completed in the spring of 1890, the new cut being opened in February.

By the end of the 1880s both Union canals were in poor condition. The Grand Junction, prompted by Fellows, Morton & Clayton (FMC), the main carriers on the route, inspected both canals and suggested means to encourage through traffic. The companies replied offering to sell out, whereupon the Grand Junction suggested £5,000. The companies countered with a request for £25,000, as they considered they had a major asset in their water supplies. This was too much for the Grand Junction, and the matter lapsed.

In February 1893 it was suggested that FMC could show more enterprise in trying to attract trade to the route. A meeting followed at which Joshua Fellows, one of FMC's directors, pointed out that trade was almost impossible at present, but if both canals were thoroughly dredged, and the locks at Watford and Foxton widened, his company would put on large steamers to compete with other forms of transport.

Fellows went on to suggest the Grand Junction should buy both Union canals and modernise the Leicester line. The two companies now asked for £20,000. Fellows was then authorised to open negotiations, acting as the Grand Junction's agent. On 12 July 1893 he reported he had obtained the GUC for £10,500 and the L&NUC for £6,500, with £250 to be paid in compensating their joint clerk. Later the price was increased by £150 to include winding up costs. The companies held winding up meetings on 16 November 1893, and the transfer Act[21] was passed in 1894, the actual purchase taking place on 29 September.

The Grand Junction immediately began a comprehensive dredging programme which continued for the next ten years. It also opened negotiations with the Leicester and Loughborough Navigations and the Erewash Canal for a reduced through toll. The Grand Junction agreed to guarantee £350 to the Leicester, £200 to the Loughborough, and £150 to the Erewash if the through tolls did not reach these minimum figures each year. This gave it virtual control over the tolls on the waterways line from the Derbyshire

FMC was an important carrier to Leicester. Here the steamer *Earl* is at the public wharf at Memory Lane, Belgrave, surrounded by barrels of glucose. *John Anderson*

For dredging the canal the Grand Junction acquired a grab dredger from Grafton & Company of Bedford in 1896, and many more followed in later years. The Grafton could with its boom position the grab with considerable accuracy, and the steam cylinder gave a positive action to the jaws. Stability was achieved by attaching side pontoons which added another 7ft. to the 7ft. of the dredger hull; by detaching the pontoons the dredger could pass through narrow locks.

and Nottinghamshire collieries to London. Early in 1897 it also secured options to purchase the three navigations: the Leicester for £38,075, the Loughborough for £26,250 and the Erewash for £21,000.

At the same time, and spurred on by FMC, plans were drawn up to widen Watford and Foxton locks. The carriers were certain that with wide boats in operation they could reduce their costs and increase traffic between London, Leicester, Nottingham and Derby.

Action was certainly needed. Whereas 125,000 tons of Derbyshire coal had reached the Grand Junction in 1805, this had dropped to 73,000 tons in 1855, and to a mere 4,700 tons in 1894,

To test the feasibility of the proposed design for Foxton, an experimental inclined plane was constructed by the Grand Junction near its Bulbourne depot in 1896. *National Waterways Museum*

the decline reflecting intense competition for the lucrative traffic, particularly from the Midland Railway.

There was considerable support for the widening proposals, and in July 1896 the company agreed to construct an inclined plane at Foxton to bypass the ten narrow locks. It would be able to take a seventy-ton barge, or a pair of narrow boats. A decision on Watford locks was deferred, but FMC agreed to put on extra narrow boats whilst the plane was being built.

The development of the plane was entrusted to Gordon Thomas, who had originally suggested it. With the help of his brother James he constructed a large scale model, powered by a portable steam engine, at Bulbourne works to test the feasibility of the design. It consisted of two watertight tanks, resting level on carriages, which ran on rails up and down the incline, counterbalancing one another.

In November 1896 a large party of experts visited the trial installation, and shortly afterwards the company invited tenders for the full-scale work at Foxton. A year later the tender of J.& H. Gwynne & Company of Hammersmith was accepted for the supply and erection of the steelwork for £14,130. Construction started early in 1898 with the company carrying out the earthworks for the plane and the two canal arms, together with the brickwork, by direct labour. At the height of the activity between 200 and 300 men were employed.

The plane was completed in June 1900 and opened to traffic on 10 July, the first boat to pass through being the steam inspection launch *Gadfly* used by Thomas Millner, the Northern district engineer. The first working boat through was reported to have been FMC's steamer *Phoenix*.

The installation at Foxton comprised two parallel sloping planes with a tank running sideways on each. The tanks were 80ft long, 15ft wide and 5ft deep inside. When filled with water they weighed about 250 tons, each being supported on eight pairs of wheels, running on rails arranged in four pairs. As the boats were

Opposite page, top: A pair of FMC horse boats entering one of the tanks of the Foxton lift in July 1900 during a visit by the company's directors on the day after the formal opening. *John Anderson*

Opposite page, bottom: The winding drum of the Foxton lift. *The Thomas Lift booklet.*

carried afloat the weight was the same, whether they were loaded or not.

The tanks counterbalanced each other and were linked by a wire balance rope, secured to the centre on each side, which passed round pulleys at top and bottom of the plane, thus forming a continuous circle. There were also two 7in steel wire haulage ropes attached to the ends of each tank, and passing round winding drums in the engine house at the top. These ropes were carried on rollers set into the face of the plane.

To permit staggered berths at top and bottom the northernmost plane was set back somewhat. At the bottom the tanks were submerged and the boats simply floated in or out, but to compensate for the loss of counterbalancing weight as the descending tank submerged, the gradient of the rails at the top was reduced by making them curve over. To prevent the tanks from tilting, however, an additional pair of leading wheels on a short upper leg came into operation on short sections of track which continued the gradient.

Guillotine gates were fitted at each end of the tanks and at the two ends of canal at the top of the plane. When at the top, the tank was sealed to the end of the canal by hydraulic pressure applied through buffers at the other end. The winding drums were driven through worm gearing by a steam engine, which also worked the hydraulic pump for the buffers and the guillotine gates[22].

The plane overcame the 75ft change of level and was 307ft long with a 1 in 4 gradient. Officially it could pass a boat in twelve minutes,

Below: The Foxton plane and locks.

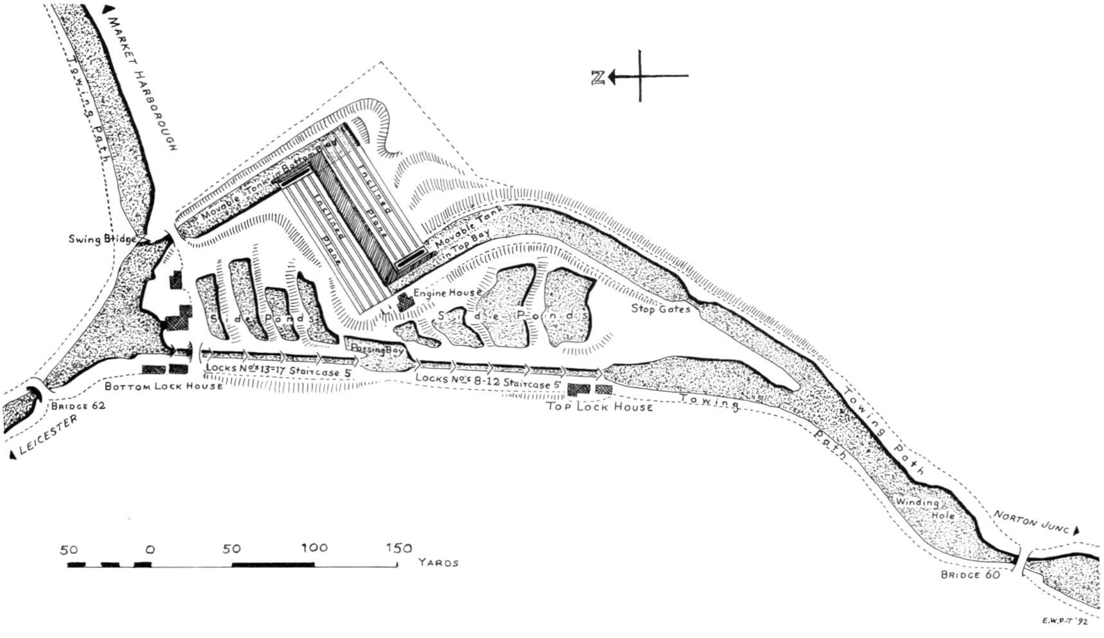

but in practice times of about five minutes could be achieved for boats in a hurry, compared with seventy minutes for the narrow locks. Another benefit was the saving in lockage water.

In the long run the plane was not successful. There were subsidence problems, with the rails giving way and being distorted

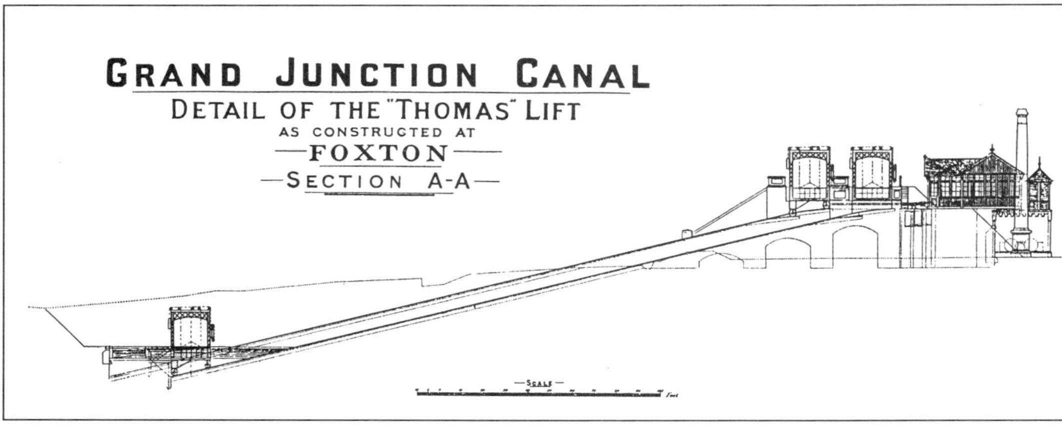

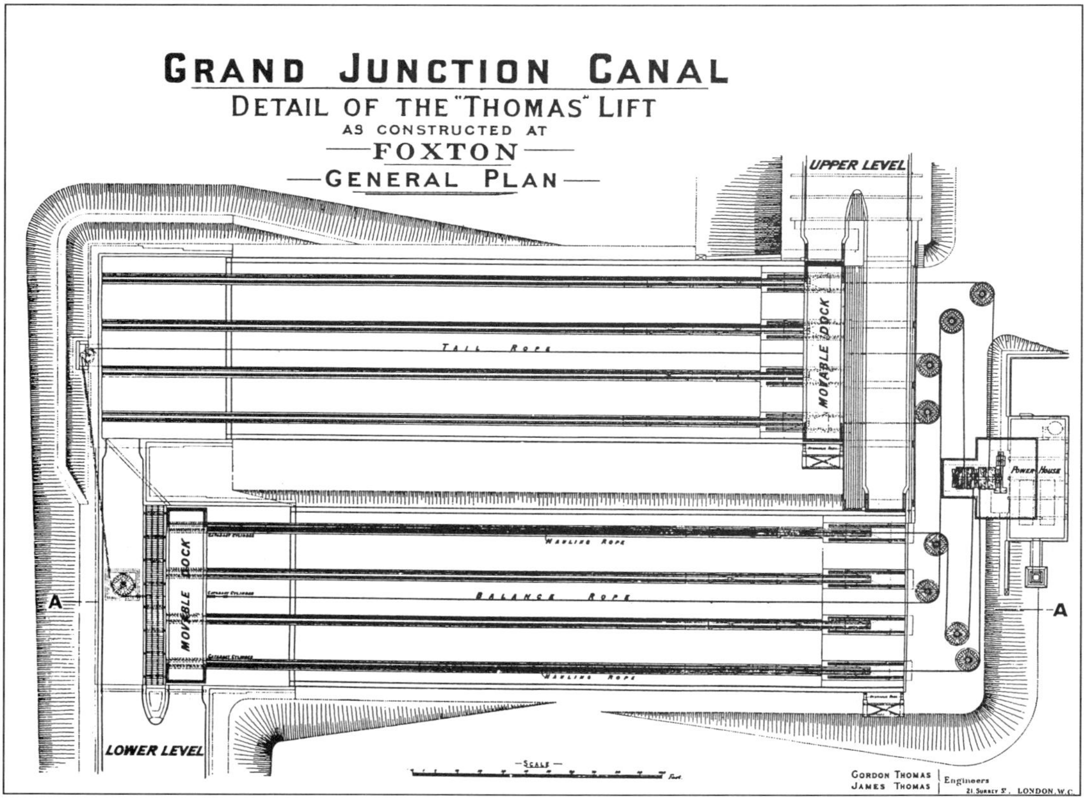

by the weight of the tanks. More significantly traffic over the Leicester line was too irregular to justify keeping the engine in steam, and paying the staff of three—one in the engine house and attendants at top and bottom to operate the guillotine gates and supervise boats entering and leaving the tanks. The capacity of the plane was about 3,000 tons in each direction in a twelve-hour shift, but the traffic never approached these figures.

A similar plane was proposed to bypass the seven narrow locks at Watford, but even before Foxton was working the company decided against it, probably due to the initial cost and running expenses. In February 1900 the widening of these locks was authorised at a cost of £17,000, but then postponed until Foxton was working satisfactorily.

The company was becoming disillusioned, for there had been little increase in through traffic, and payments were having to be made regularly under the guarantees to the Leicester, Loughborough and Erewash companies. In 1900 the purchase options were allowed to lapse, but in the following year the three concerns offered themselves for £50,000. The Grand Junction declined, pointing out that whilst it was disappointed at having to

Opposite page, top: A cross-section of the Foxton lift.

Opposite page, bottom: General plan of the Foxton lift.

Below: Discharging dredgings from mud hoppers near Yelvertoft in 1896 during the comprehensive restoration work carried out by the Grand Junction after it had acquired the two Leicester canals. *Hugh McKnight*

give up a scheme to develop traffic, it had already spent a large sum for no return. It refused again in 1902, fearing the three navigations would become as much a burden as the two Union canals were proving to be.

The widening experiment finished when the Watford locks were rebuilt as a narrow flight at a cost of nearly £10,000 between November 1901 and February 1902. In June 1902 the company announced that the extensive dredging programme from Norton to Leicester was virtually complete.

Gordon Thomas claimed that through trade had failed to develop due to the poor state of both the Cromford Canal, owned by the Midland Railway, and the Erewash Canal. The high tolls on the Nottingham Canal, controlled by the Great Northern Railway, and the disuse of the Nutbrook Canal were other contributory factors[23].

The Foxton plane continued to operate, but it was only a matter of time before costs dictated it should be closed. In November 1908 it was decided to reinstate Foxton locks to pass traffic at night, when

Work started on the reconstruction of Watford locks on 11 November 1901. Traffic was stopped from 13 November 1901 until 21 January 1902, and again from 4 February to 7 February. At one time 160 men were engaged. The work cost £5,545, of which £2,846 was for labour, but this excluded the new lock gates that were supplied from the company's works at Bulbourne, near Tring. *National Waterways Museum*

the plane was not working, and this was done during 1909. On 26 October 1910 it was announced that the plane would cease working in a fortnight, and all traffic was then to use the locks. The plane was used occasionally after this, but in 1914 part was dismantled, and the rest was taken down between 1924 and 1926. Some of the material was re-used by the company; the remainder was sold for scrap in 1928[24]

A detail of one of the tanks of the Foxton lift.

With the passage of time much of the plane has become overgrown, but a Trust was founded in 1980 to work towards its long-term restoration. A museum has been created in the old engine house, and further works are planned. The short branch to the Old Union at the bottom of the plane is now used for moorings, but the branch at the top is completely overgrown.

This ended the Grand Junction's attempt to improve the Leicester line. The two canals had cost £19,123 to purchase, and by 1928 almost £78,700[25] had been spent on reinstating the works. This included the plane, which cost almost £40,000. To finance the work money was raised on 4% debentures, but with only a marginal increase in traffic it proved an expensive venture.

NOTES TO CHAPTER EIGHT

1 Hadfield, Charles. *Canals of the East Midlands,* 1966.
2 Loughborough Navigation Acts: 6 Geo III, c 94—14 May 1766 & 16 Geo III, c 65—2 April 1776.
3 Erewash Canal Act: 17 Geo III, c 69—30 April 1777.
4 Leicester Navigation Act: 31 Geo III, c 65—13 May 1791.
5 *Leicester Journal,* 23 March 1792.
6 *Derby Mercury,* 14 June 1792.
7 *British Chronicle,* 5 December 1792.
8 33 Geo III, c 98—30 April 1793.
9 *Leicester Journal,* 31 October 1794.
10 Stevens, Philip A. *The Leicester Line,* 1972. (This contains a detailed history of the Old Union and Grand Union Canals).
11 Reports by Jessop and Barnes in the Library of the Institution of Civil Engineers.
12 45 Geo III, c 71—27 June 1805.
13 *Northampton Mercury,* 21 October 1809.
14 *Leicester Journal,* 3 June 1808.
15 *Northampton Mercury,* 17 December 1808.
16 50 Geo III, c 122—24 May 1810.
17 *The Leicester Line,* op cit.
18 Ibid.
19 *Derby Mercury,* 18 August 1814.
20 44 & 45 Vic, c 72—27 June 1881.
21 57 & 58 Vic, c 85—20 July 1894.
22 *Engineering,* 25 January 1900 (illustrated account of the plane).
23 Evidence to Royal Commission on Canals & Inland Waterways, 1906.
24 Grand Junction published accounts, 1925–28.
25 Ibid, 1928.

SOUTHWOLD

The End of an Era 9

THE WORK ON the Leicester line was not the only improvement carried out by the company in the last quarter of the nineteenth century. In February 1878 Hubert Thomas suggested building a branch canal from Bulls Bridge to Slough, which would open up new brickfields to supply the expanding London market[1]. He was authorised to carry out a preliminary survey and make an estimate of the costs. In July the company's solicitors were instructed to apply for Parliamentary powers, and by the end of October a Bill had been drafted. This was approved at a special meeting of the shareholders in January 1879, when Thomas reported the results of his survey. His estimate was £70,550, but the junction with the main line was now to be at Cowley Peachey.

The Bill was opposed by the Duke of Northumberland, who was concerned about the effect of the branch on the river Colne water supplies, and by the Regent's Canal Company, which was anxious about the adverse effect on the water supplies to the Long Level. Protective clauses were inserted, resulting in the withdrawal of these petitions. Other opponents were the GWR, as the new branch was to run parallel with its line, and landowners in the Langley area. Despite the opposition, the Bill passed through the House of Lords committee on 26 March and went to the Commons. Here the committee hearing started on 3 July, and after four days the Bill was approved, but the company had to agree to terminate the branch at the Stoke Poges road in Slough rather than further west as had been planned, to avoid interference with land belonging to Eton College and to the Duke of Leeds. The Act[2] was passed in July.

Preparatory work then started and a sub-committee was set up to oversee matters. The monies needed were to be raised by mortgage. Construction started in earnest early in 1880 at the eastern end, where the heaviest works were concentrated. In the first 1,300yd there were aqueducts over Fray's river, the Colne, and Colne Brook, and part of the canal was on a substantial embankment. At Iver a deep and lengthy cutting was required, and nearer Langley more heavy cutting and embanking works were needed.

By the middle of 1881, when the eastern section was progressing well, some of the labour force was diverted to the Slough end, where the works were comparatively easy. On 1 December 1881 Thomas, accompanied by other company officers, navigated the branch from the Iver road bridge to the main line in a steam launch. Iver cutting, which cost over £10,000 in labourers' wages alone during the fifteen months it was being excavated[3], was finished about April 1882. The branch was opened as far as Langley by the end of May, and on 4 December 1882 Thomas reported it was open to traffic throughout[1].

Opposite page: Sabey's wide boat *Southwold* being unloaded at Willesden electricity works, which relied heavily on canal transport, with coal being delivered from Warwickshire collieries by narrow boat. Ash from the works was also taken away by canal.
National Waterways Museum

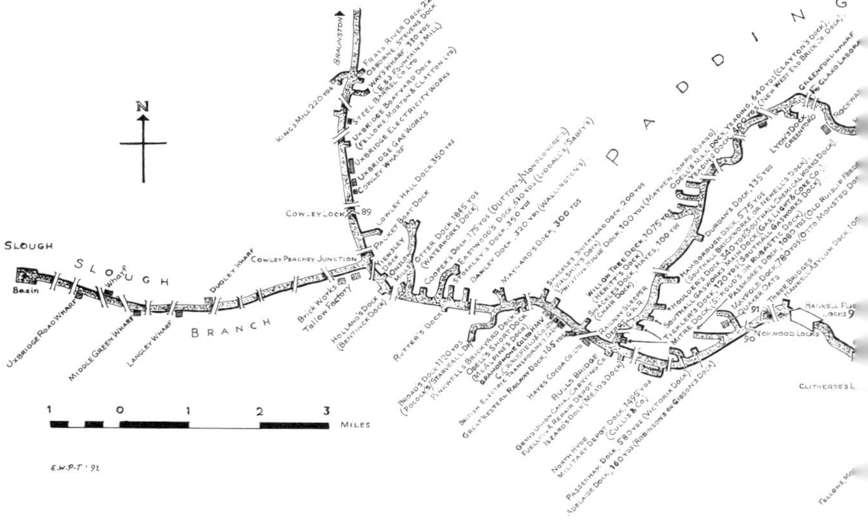

The complex of short docks, and some of the wharves, that were centred on Bulls Bridge.

The opening precipitated a dispute with the Regent's Canal Company about water supplies, which dragged on for many years. On the positive side the branch gave a major boost to the brickmaking industry and many wharves were established, particularly in the Iver and Langley areas. Alongside the brickworks, sand and gravel excavation was developed, and this provided the main traffic when the supplies of brick earth started to run out.

The tonnages on the branch were substantial, totalling 192,200 in 1905. They had declined to 122,060 in 1915, and to 103,859 in

A wide boat, well loaded with bricks destined for the London area, above Harefield's Copper Mill lock in about 1895. *Chris Herridge*

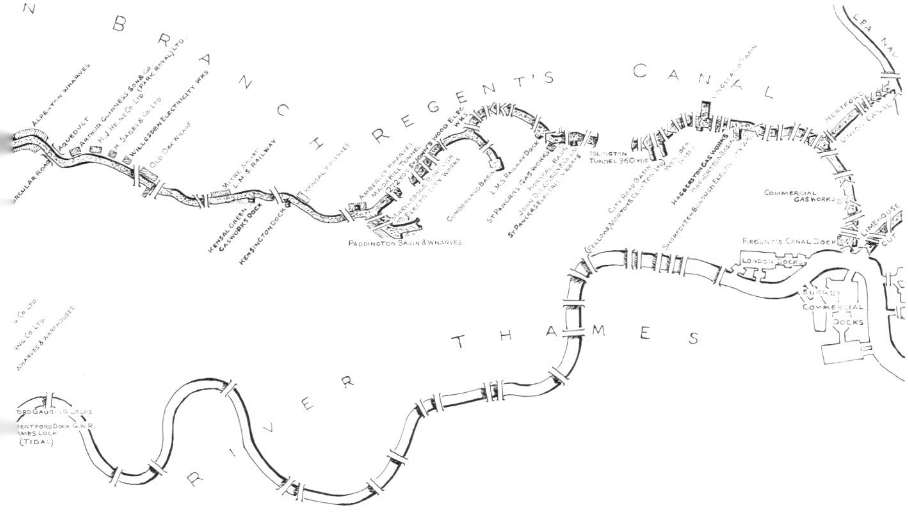

1927[5]. Tolls yielded £7,614 in 1905, and £3,872 in 1915. As maintenance costs were low, the branch made a useful contribution to profits.

Another development was on the short Rickmansworth branch, where a basin and wharf were built in 1871 alongside the railway line from Watford, to provide transhipment facilities. In 1903 a side lock was built, linking up with another section of the river Chess, to enable sand and gravel extracted from there to be carried away by canal.

Several other pits in the area were also linked to the main canal to permit gravel to be transported out by barge. In 1913 the Rickmansworth Gravel Company made a connection to Stockers Lake, just south of Rickmansworth, and in the same year a cut was made to Batchworth Lake.

A network of short branches, or docks, was created in the Yiewsley and Hayes area to penetrate into brickfields and provide a ready means of transport for both raw materials and the finished products. The output of bricks from these yards, and from those on the Slough branch, went to supply part of the insatiable demand created by the continuing expansion of London at the time.

At Yiewsley Liddall's Dock was extended by Eastwood & Sons in 1881, and again in 1913 by H. Sabey & Company, to over 600yd to link up with some ballast pits. At one time a tramway ran from the end of the dock further into the pits, and the 1913 extension was probably over the site of the line, to avoid transhipment. In 1904 William Boyer & Sons built a new dock nearby, known as Stockley Dock, and lengthened it to 350yd in 1914.

Similarly Pocock's Dock was lengthened in 1880, and again in 1895, to over 1,100yd and a tramway extended from its south-western

A steel lift bridge on the Rickmansworth branch.

Opposite page: The ice boat *Paddington,* built by Walkers' at Rickmansworth in 1919 for the Grand Junction Canal Company, at work breaking a channel through thick ice. Such icebreakers were effective because as they were towed ahead by a large team of horses they were rocked by men tugging at the midships bar so as to make as wide a path as possible.

end further into the brickfield. This dock, crossed by the GWR and by the Yiewsley to West Drayton road, was eventually taken over by Broad & Company.

On the Paddington branch a dock was built in 1853 into a brickfield at Yeading; known as Yeading Dock, it was 400yd long. Another was built nearby in 1860 by Henry Dodd, and was later taken over by Odell & Company. Another at Yeading was known as Willow Tree Dock; it was eventually over 1,000yd long. The Hamborough Dock at Southall was built in 1863, and later extended to nearly 600yd. In 1869 a dock was opened into the new Southall gasworks, with a second in 1909. The nearby Stroud's Dock was built in 1870, passed to the International Tea Co. Ltd. in 1915, and remained in use until 1970.

Nearer Paddington the L&NWR built a basin at Willesden in 1855 to facilitate transhipment from railway wagons to barges, so as to reach destinations on the Regent's Canal. Later coal drops were erected there, and it became known as Mitre Tip. The basin was enlarged in 1885. In 1851 the Western Gas Light Company built a dock at Kensal Green to serve its works, and a second basin was opened in 1883. Nearby Kensington Dock was opened in 1894 to serve a rubbish collection depot for the Royal Borough of Kensington.

On the main canal a new dock nearly half a mile long was opened in December 1913, running from just above Norwood top lock, to serve the Maypole margarine works. Known as the Otto Monsted or Maypole Dock, it cost £27,670, and was constructed with concrete piled banks. A substantial traffic quickly built up in coconut oil, margarine and other foodstuffs. Almost 55,000 tons were carried in 1914, and this rose to a peak of over 84,500 tons in 1920[6]. Thereafter it declined, but was still over 30,000 tons in 1928.

The last of the docks to be opened was in 1922 with a short branch into the Greenford factory operated by J. Lyons & Co. Ltd. on the Paddington branch. This received its supplies by barge through Brentford, and traffic built up quickly. Nearly 10,000 tons was handled in 1922, almost 19,000 two years later, and over 15,500 in 1928[7].

As some docks opened, others closed. Cowley Hall Dock closed in 1893, and at Yiewsley most of the long Otter Dock was filled in between 1908 and 1910, leaving only a short section to allow coal to be delivered to a waterworks pumping station. Passmore's Dock was partly infilled in 1895, and by 1925 most of the rest had followed. Then as the brickfields were worked out, their docks were used to receive rubbish carried by barges along the Long Level from London boroughs such as Marylebone and Paddington.

An 1836 plan to link the western end of the North Hyde Military Dock with the main canal and complete the loop, as originally intended, was not carried out. Neither was a plan in 1847 for a 320yd branch canal southwards from the dock to open up a brickfield[8]. Instead two tramways were built to wharves on the dock,

Ironbridge lock, No 77, at Watford was said to be haunted by the ghost of a coloured servant from a nearby house who did his utmost to hinder the passage of boats.

where transhipment to barges took place. Later the far end of the dock became disused, and this had been filled in by 1920. The rest followed in 1935, once the brickfield was worked out.

The company did all it could to keep its system competitive. Traffic was still at a high level, despite intense railway competition and the growing impact of road transport. Total tonnage was 1,172,463 in 1888, 1,602,552 in 1898, and 1,794,233 in 1905[9], but by 1925 had declined to 1,485,346[10].

The peak traffic was in 1914, when 348,470 tons passed through Marsworth top lock 45, with 34,368 tons in 2,002 boats in May alone. Likewise 440,516 tons passed through Norwood lock 90, with 46,163 tons in July alone[11].

The number of craft operating was also substantial: 6,790 pairs and 1,555 single boats passed through Braunston locks in 1908, carrying some 295,000 tons. In the same year 7,466 pairs and 1,570 single boats passed Stoke Bruerne top lock, whilst 3,299 boats passed along the Northampton branch. There was also a fair proportion of empty running: 2,790 northbound and 512 southbound boats passed light through Braunston in 1908, whilst 3,064 northbound and 812 southbound boats passed light through lock 45, indicating the predominance of southbound cargoes.

There were several developments in the Northern district in the early 1900s. In 1893 the upper part of the north portal of Blisworth tunnel started to slip outwards on the towpath side. Temporary wooden struts were erected, but between December 1902 and October 1903 the whole portal was rebuilt in blue brickwork at a cost of £639[12]. In the early 1920s repairs to Wolverton aqueduct became necessary, and the whole of the abutments were refaced, again in blue brickwork, costing £1,700.

There was also a problem at Mountmill, on the Buckingham branch, some three and a half miles up the arm. At this point the canal runs close to the Great Ouse, but at some height above river level. For some time leaks had occurred here, and in 1919 matters became so bad that the affected stretch had to be drained and the bottom and sides concreted for about 200yd to cure the trouble.

Another problem on this branch was at the extreme upper end in the town. This section did not belong to the company, and into it Buckingham Corporation discharged all the town sewage. Refuse from a feeder connecting with the river Great Ouse also accumulated in the canal. Periodically the company cleaned the branch out, on occasion going right into the town, but only legal action at the end of the 1890s, when an injunction was obtained against the Corporation, finally put an end to the trouble[13].

By then the damage had been done. In the early 1890s 2,500 tons of coal and 500 tons of other goods reached Buckingham each year, but the final section was in such poor condition that traffic was falling off. In the 1900s thirty boats a year were still going up to Maids Moreton mill, one and a half miles below Buckingham, with roadstone for the local council, but trade to the town itself had ceased. Roadstone was also being taken to Leckhampstead wharf, and there was some agricultural traffic, and some trade to Deanshanger[14]. In evidence to the Royal Commission in 1908 Buckingham Corporation promised to take 500 tons of stone and 60 tons of coal by water if the branch was re-opened, but the offer was not taken up.

The biggest problem of all was on the Tring summit, due to continuing leaks from the Wendover branch. During the summer of 1897, a particularly dry season, the leakage was so great that it was more than the entire supply coming into the branch at Wendover, and was drawing water from the main line, making it impossible to maintain a depth of more than 4ft on the summit. On 21 August two sets of stop planks were put in at Little Tring, and the intervening space pumped out, thus severing the branch from the main canal[15]. The main line water level immediately improved, whilst on the other side of the dam the level sank, notwithstanding the supplies coming in from Wendover.

The portal of Husbands Bosworth tunnel, on the Leicester line to the south of Foxton. The old horse path across the top of the tunnel can still be walked.

In 1898 this temporary arrangement was made permanent with the construction of a stop lock. The intention was for the branch to be available when there was sufficient water, but the dry weather persisted and the branch effectively remained closed, except for a short time in 1901 when a few boats navigated up to Buckland. Whilst the branch was closed the Wendover water was run down into Wilstone reservoir.

Several complaints were made about the closure. In 1903 Lord Rothschild and other landowners approached the Railway & Canal Traffic Commissioners, who decided the Act that authorised making the branch navigable was permissive, and not obligatory. The only obligation was to make the branch a feeder. For its part, the company claimed that keeping the branch open would merely endanger the main canal.

In April 1903 the surviving stub of the branch from Tringford to the main line and the main line from the junction to Bulbourne bridge was drained and repairs carried out to reduce the leaks. On the main line this involved building brick walls on both sides and lining the bed with concrete and bitumen sheeting.

Gordon Thomas then made three suggestions for the closed branch[16]. The first was to restore the navigation by repairing the asphalt and puddle at a cost of nearly £15,700; the second, costing £19,500, was to repair the asphalt and puddle and provide brick and concrete side walls between Little Tring and Drayton; the third was to convert the branch into a water conduit, at a cost of almost £4,900, with the water level being permanently lowered.

A variation of the cheapest scheme was carried out in 1904, when the bed of the branch between Wendover and Drayton was repaired, using clay puddle from the sides of the canal. The water level was lowered and diverted into Wilstone by the Drayton feeder, and the canal between Drayton and the stop lock was then abandoned.

This meant that the bulk of the pumping now had to be from

HASTY
EMANUEL SMITH
Canal Carrier
BRENTFORD.

Wilstone, but as this was low-lying, the costs were formidable. In September 1910 Thomas suggested piping the Wendover water along the abandoned stretch of the branch to the Tringford station, where a new pump would be installed[17]. The estimate was almost £3,500, and in November his plan was authorised.

In practice it was found difficult to design a pump to cope satisfactorily with the very variable flow of water from Wendover, which ranged from fourteen to forty-two locks per day. In April 1911 Thomas proposed instead to pass the Wendover water into a mixing shaft at the pumping station, where it could either be pumped by a new electric pump into the summit or passed by gravity to Tringford reservoir. He also proposed reducing the size of the pipeline, which would save £1,000, scrapping the old York engine, which was nearly worn out, and replacing it with a new electric pump. Electricity was to be provided by two diesel-driven generators.

The cost of all these improvements was about £8,000, but a saving of 4½p per lock was expected in pumping costs, due to the much higher proportion of water to be drawn from the high-level Tringford reservoir. A contract for nearly £4,000 for the new pumps and electrical equipment was awarded to Rees Roturbo Ltd. in July 1911. The Grand Junction carried out all the other works and the pipeline was finished in the early summer of 1912[18]. A connection was made to the old heading at Whitehouses so that water could be diverted to Wilstone if required. On 12 June 1912 the water was turned on, and the new pumps were completed a year later.

Soon afterwards it was decided the Old engine was due for renewal; subsequently it was only used occasionally to supplement the other pumps. In 1926 William Yates, the engineer, recommended its replacement by new electric pumps and his plan was accepted in May. Two Rees Roturbo pumps were supplied and placed at different heights in the pumping well. Both were needed to provide the lift from Wilstone, but either could draw from the high level reservoirs. The new plant, opened on 19 October 1927, cost £4,292 of which £2,344 was for the plant and the rest for the constructional work, which included dismantling the historic beam engine. Electricity was supplied from the Borough of Aylesbury power station.

In the summer of 1902 a new engine and pump were installed at Cowroast, improving the supply from the well significantly. There had been a series of dry years since 1895, culminating in a severe drought in 1902. All the reservoirs were virtually empty, and it was only the newly equipped station at Cowroast that enabled the company to keep the summit open at all. Even then there were long delays, as there was only enough water for up to ninety pairs of boats a week[19], whereas normally 130 would pass over the summit.

Throughout the 1880s and 1890s the company held its dividend at a regular 4%, but this fell to 3% in 1903. In 1918 it started climbing back up to 4% again. A decreasing proportion of income came from tolls, and it was the revenue for the many properties along the canal, and especially at Paddington, that enabled a dividend to be

Opposite page: Brentford, the Grand Junction's outlet to the Thames, developed into a centre for transhipment between lighters and other river craft and narrow boats, as lighterage traffic rarely went further up the canal than Berkhamsted. Imported timber became a principal cargo, remaining such until the end of narrow boat carrying. The motor boat winding is Emanuel Smith's *Hasty* of 1912, the second motor built by Walkers' for the Brentford carrier.

paid, as toll receipts had been fairly static since the 1870s.

The figures for 1928 were typical with tolls at £71,375, rental income at £43,156, and the Paddington income another £27,360. Canal expenses were £48,354, and other expenses £47,820[20]. This left nearly £56,000 available, of which £6,000 was needed to pay the 4% interest on the £150,000 debenture stock, and £5,622 the 6% dividend on £93,700 preference stock. Most of the remainder went to pay a dividend on the £1,130,000 ordinary capital, which absorbed £45,200 at 4%.

The company continued to spend out to modernise the canal. At the turn of the century much work was carried out at Brentford, the rural area around the lock being developed with the building of extensive wharves and warehouses. The lock itself was duplicated in 1900 to handle the increased traffic, but plans for a second Thames lock and a new channel up to Brentford bridge were abandoned, as Parliamentary powers would have been needed.

An extensive dredging programme was carried out on the main line and on the Paddington branch. This originated with the idea of allowing 14ft beam craft to work through to Leicester, and the company invested heavily in new steam dredging plant for this work.

In 1925 improvements were carried out to Walton mill in

The first part of a factory to serve the H. J. Heinz Company was built at Harlesden alongside the Paddington branch in 1925, and the company depended on canal transport until well into the 1960s. Here a barge lies in the unloading bay; it might well have been carrying beans, which was one of the last traffics handled by water. *National Waterways Museum*

Aylesbury costing £2,400; in 1926 a new warehouse was provided at Northampton wharf costing over £1,250; and in 1927 a new electric crane and warehouse improvements at Brentford cost nearly £1,600. In addition much walling was carried out in the Brentford area to increase the cross section of the canal, and to protect the banks. On the income side a block of land at Paddington was sold to the Governors of St Mary's Hospital in 1927 for £65,000[21].

The fairly static toll income was partly due to the emergence of FMC as the largest carriers on the canal. Its size enabled it to obtain fine rates, and the Grand Junction was very dependent on this toll income. During a protracted strike by FMC's boatmen in the autumn of 1923, Yates had to cut back heavily on maintenance expenditure until trade was back to normal.

In April 1888 Winthrop Mackworth Praed was elected the company's chairman, taking over from the Hon Robert Howe Browne. Praed died in February 1890 and John Stone Wigg was appointed. He died in December 1898, when Thomas Tatham succeeded him. Tatham served until his death in March 1914, when Rodolph Fane de Salis took over the chair, which he then held for the remainder of the company's independent life. He had joined the select committee in June 1888, and was also a director of the Coventry Canal.

Rodolph Fane de Salis, born in 1854, served as the Grand Junction's last chairman from 11 March 1914 to 31 December 1928. He died on 26 November 1931 at the age of 77. In his time he also served as President of the Canal Association, as a director of the Coventry Canal, and of Singer & Co of Coventry. *National Waterways Museum*

In 1905 the clerk and general manager, Hubert Thomas, was in poor health and retired. He was succeeded by John Bliss, who had started as a junior clerk in 1877, and remained until the end of 1928. Hubert's son Gordon served as the company's engineer until he was dismissed in 1916 for suspected embezzlement. He was replaced by William Yates.

The First World War caused many problems for the company, as skilled men joined the armed forces or were attracted to higher paid work in the armaments factories. The carriers were similarly affected, and traffic levels quickly dropped. Meanwhile toll rates had been frozen. To arrest the decline, the canal was brought under Government control from 1 March 1917. Compensation was then paid for loss of profits, which enabled the company to retain its workforce by paying a higher wage.

The compensation was based on the company's accounts to 31 December 1913, which recorded a net profit of £45,821, based on 1,237,121 tons handled[22]. In 1917 tonnage was only 1,090,138, and the company received £23,764 for the ten months it was under Government control. The compensation rose sharply in both 1918 and 1919, despite some recovery in traffic. Government control came to an end on 31 August 1920, when a major increase in tolls was authorised for both the canals and the railways.

After the war thoughts on widening the company's activities were resumed, but this time the Birmingham line was studied. In June 1925 Yates reported on the cost of putting the three Warwick canals that made up the through route into a proper state of repair. An attempt to take them over had been made soon after the Union

Left: Towards the end of the Grand Junction era Brentford grew in importance. This busy scene was typical, with timber being transhipped from barges to narrow boats for carrying to the Midlands. *National Waterways Museum*

Opposite page: Two maintenance boats pass the entrance to Cooper's Dock at West Drayton in 1959. This was one of several such docks in the area. *Lewis Edwards*

Below: The Brentford area, showing the Canal Boatman's Institute opened in 1904, the Great Western's Brentford Dock and other features.

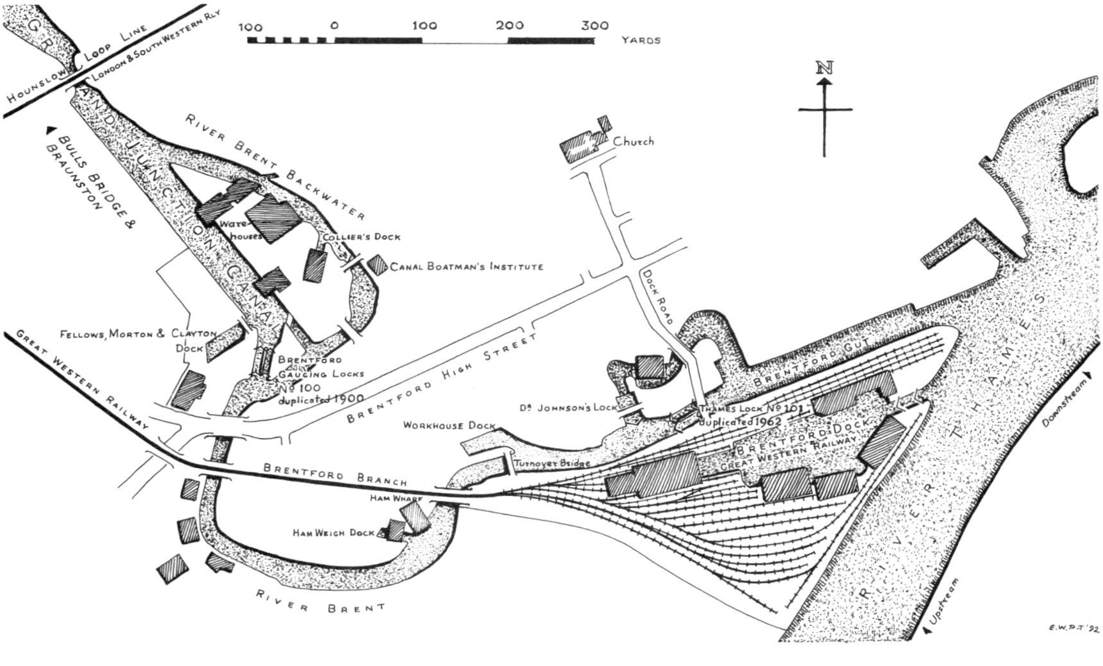

canals were acquired, but the enabling Bill had to be withdrawn from Parliament in June 1895 because of opposition from local authorities on rating grounds.

A committee was set up to investigate a possible purchase or merger. This soon involved the Regent's Canal & Dock Company whose chairman, Wilfred Curtis, was also on the Grand Junction's select committee. The Regent's suggested it should purchase the Grand Junction, excluding its interest in the Paddington estate. The idea was welcomed, and a provisional agreement was reached in November 1926 whereby the Grand Junction was to purchase the Warwick canals, which were being managed by a joint committee at the time.

By the end of 1927 the details had been finalised, although for technical reasons it was decided the Regent's should purchase the Warwick canals direct, rather than through the Grand Junction. Bills were deposited in January 1928 to authorise the takeovers.

The Acts[23] received the Royal Assent in August, and from 1 January 1929 the combined concern came into being as the Grand Union Canal Co. Ltd., a title adopted by the Regent's to signify its wider interests.

The Grand Junction transferred its entire canal assets to the new company, receiving £285,709 5½% debentures in return. It then promoted a further Act[24] to convert itself into a limited company. This remained in being as an independent property concern until the end of 1971, when it was taken over for nearly £28 million to

The bridge at Little Tring over part of the dry section of the Wendover branch, seen in May 1965. It has since been demolished. In 1989 a Wendover Arm Trust was formed to work for the re-opening of the branch, and the removal of the embankment which has replaced the bridge is one of its first objectives.
Alan Faulkner

join the Amalgamated Investment & Property Co. Ltd. Meanwhile the Grand Junction shareholders received Grand Union ordinary shares in the ratio of £67.30 for every £100 of Grand Junction stock.

The Grand Junction held its final meeting on 12 December 1928, when compensation for the retiring select committee and the retiring clerk was fixed, and a portrait of de Salis was presented to him as a parting gift. It was the end of an era which had lasted nearly 140 years, and covered a significant period in the history of this country. From 1929 the Grand Junction carried on, but as a part of a much larger concern.

NOTES TO CHAPTER NINE

1 Select Committee minutes, 27 February 1878.
2 42 & 43 Vic, c 178—21 July 1879.
3 Slough Branch Canal cash book, 1879–82.
4 Select Committee minutes, 5 December 1882.
5 Grand Junction tonnage book, 1904–28.
6 Ibid.
7 Ibid.
8 Compton, Hugh & Faulkner, Alan. North Hyde Military Dock, *Railway & Canal Historical Society Journal,* July 1992.
9 Royal Commission on Canals & Inland Waterways, 1908.
10 Cadbury, George & Dobbs, S.P. *Canals and Inland Waterways,* 1929.
11 Grand Junction tonnage book, 1904–28.
12 Photographic record book by T.W.Millner in the National Waterways Museum.
13 Royal Commission on Canals & Inland Waterways, 1908—35290 & 41381.
14 Ibid, 35302.
15 Ibid, 41357.
16 Richardson, Alan. Water Supplies to Tring Summit. *Journal of the Railway & Canal Historical Society,* April 1969.
17 McGarey, D.G. *The History of Tringford Pumping Station,* 1946.
18 Ibid.
19 *Daily Mail,* 7 October 1902.
20 Grand Junction Canal annual report, 1928.
21 Grand Junction Canal annual reports, 1923–8.
22 Canal Control Committee records, Public Record Office MT 52/4.
23 18 & 19 Geo V, c 98—3 August 1928. (Grand Junction purchase)
18 & 19 Geo V, c 99—3 August 1928. (Warwick Canals purchase)
24 20 & 21 Geo V, c 22—26 July 1929.

The Grand Union and Beyond 10

ON 1 JANUARY 1929 the Grand Junction began a new phase in its existence as an important part of the enlarged Grand Union concern. For the first time the canal route from the Thames, at both Limehouse and Brentford, to Birmingham was under unified management.

Soon afterwards the Grand Union began negotiations to acquire the Leicester and Loughborough Navigations and the Erewash Canal, and agreement was reached on 20 April 1931. The Leicester had various investments, and including the cost of these and compensation payments to the chairman, committee members and staff, the final purchase price was £48,127. The figure for the Loughborough was £11,595 and for the Erewash £15,702. The purchases were embodied in an Act[1], and from 1 January 1932 the three navigations became part of the Grand Union, which now controlled 276 miles of main canal and branches and 294 locks.

In August 1929 the company applied to embark on a major development programme, under a Government scheme designed to relieve unemployment. The main proposals were to deepen and widen parts of the canal, to widen some of the locks and bridges, to carry out walling, piling and dredging, and to provide a fleet of mechanically propelled barges.

The Government agreed to cover the interest on a £500,000 development loan, an enabling Act[2] was passed, and work started in March 1932. Most of it was on the Warwick section, to enable 100-ton barges to work through to Birmingham. Between Napton and Knowle fifty-two narrow locks were replaced by fifty-one wide ones, and eight bridges were rebuilt to provide a wider waterway. The Duke of Kent carried out a formal opening ceremony at Hatton top lock on 30 October 1934.

Some dredging was also carried out, together with many miles of concrete bank protection, but the money was insufficient to enable the dredging and the widening work to be completed. As a result it was found impractical to use barges up to Birmingham, as there were too many places where they could not have passed.

The company had also planned to widen the locks at Watford and Foxton on the Leicester line, but the Government refused its request for assistance and the scheme was dropped.

Improvements were also carried out on the Grand Junction section, where the cills of several locks and bridges were lowered to provide a 5½ft depth of water, to enable fully laden 14ft beam barges to use the canal up to Berkhamsted, and 12½ft beam craft from there up to Birmingham.

At the same time new mechanical dredging plant was purchased, and the canal dredged up to Berkhamsted to permit its use by the

The Grand Union Canal Company's seal, with the motto "Silent and sure". *National Waterways Museum*

deeper craft. By March 1932 the maximum cargo that barges could carry between Brentford and Willesden on the Paddington branch had risen from sixty-five tons to eighty-six tons. The ultimate aim was to allow 100-ton-capacity craft to operate. New concrete pile manufacturing depots were set up at Batchworth and Northampton, and many miles of bank protection work were carried out, most being provided with a continuous concrete coping.

As part of this work the company developed its own fleet of narrow boats to exploit its improved waterway (see Chapter 11). These were operated through a new subsidiary, the Grand Union Canal Carrying Co. Ltd., whilst a similar small subsidiary, the Erewash Canal Carrying Co. Ltd., was incorporated to develop carrying on the river Soar and the Erewash Canal. Meanwhile an important wharf and warehouse was provided at Sampson Road, Birmingham, and the facilities at Brentford, Northampton, and Leicester were improved to encourage traffic.

Much of the new trade derived from the company's Regent's Canal Dock at Limehouse. Shipping agents were appointed in 1932 to develop continental trade, and in 1937 a small steamer was purchased to begin regular services to Antwerp. Most of this traffic was exchanged with narrow boats in the dock.

In November 1937 a separate company, Grand Union (Shipping) Ltd., was incorporated[3]. It operated as the "Regent's Line" and expanded services to Rotterdam and other continental ports. An associated company, Grande Union Belge de Transports S.A., was based in Antwerp and a stevedoring company, Grand Union (Stevedoring and Wharfage) Ltd., established at Limehouse.

Another development was the purchase in April 1936 of Thomas Clayton (Paddington) Ltd., whose main contract was transporting rubbish for Paddington Borough Council to dumps at Yeading, Hayes and Southall.

The Second World War had a major effect on the Grand Union. As in the First World War, there was an immediate loss of staff and a drop in traffic. Most of the company's operations were brought under Government control from 1 July 1942, enabling tolls to be pegged and compensation to be paid for loss of profits. Regent's Canal Dock, the shipping and stevedoring companies, and Thomas Clayton were excluded.

The Government introduced a scheme to co-ordinate the various carriers, and to use the boats more effectively by cutting down on empty running. There was also a recruitment drive to replace crews who had left, and several all-female crews were taken on.

All this was reflected in the company's finances. A maiden dividend of 1.375% on the ordinary stock of £2,159,542 had been paid in 1929, but this then dropped steadily, with no dividend at all being paid from 1935[4]. In addition there was £400,000 of 6% preference stock, and no dividend was paid on this from 1937. The company also had £1,627,993 loans outstanding at the time, and a further £150,000 loan was raised soon after.

Revenue had fallen from £160,510 in 1929 to a low of £113,466 in 1940, with trade being badly hit, firstly by the depression and then by the looming international crisis. Thereafter it recovered, and reached £162,925 in 1943[5], enabling a 3% preference dividend to be resumed in 1942. It was restored to the full 6% in 1944, and in 1945 a 1% dividend was paid on the ordinary stock, rising to 2% in 1946.

This recovery was assisted by good profits from the shipping and stevedoring companies and Thomas Clayton, and by a reduced loss from the carrying company. In December 1945 a road haulage company, Cartwright & Paddock of Brierley Hill, was acquired, and new subsidiaries were set up to develop transport, warehousing, estates, and the company's reservoirs.

The general election that took place immediately after the Second World War brought in a Labour Government committed to a major programme of public ownership. This resulted in the Grand Union, along with most of the country's railways and canals, being nationalised from 1 January 1948. Most of the canal now came under the jurisdiction of the South Eastern Division of the Docks & Inland Waterways Executive of the British Transport Commission, but the part north of Leicester was in the North Eastern Division. The Executive, which traded as British Waterways,

Barges rarely worked north of Berkhamsted. Here Alfred Buck's horsedrawn barge *Trent* is on its way up to the town with timber in the late 1940s. The towline has been allowed to sink into the water to allow an oncoming narrow boat to pass over it. *National Waterways Museum*

Work in hand on building the duplicate Thames lock in 1961. The original lock on the right was later rebuilt and enlarged.
Alan Faulkner

inherited the Grand Union's narrow boats and soon acquired FMC's fleet, making it the dominant long-distance carrier on the canal.

In October 1953 the Executive was replaced by a new Board of Management, but on 1 January 1955 a more fundamental change took place. The docks were separated from the inland waterways, and British Transport Waterways was established with a general manager in charge. This followed an internal enquiry, known as the Board of Survey, set up in April 1954 by the British Transport Commission to examine the future of its waterways.

The enquiry reported in November 1954, and its recommendations for the Grand Union fell into three categories. The canal from Limehouse and Brentford up to Berkhamsted was well used and was to be developed. In contrast, the Aylesbury branch and the Erewash Canal, both of which carried virtually no traffic, were to be closed. The rest of the canal, which carried some traffic, was to be retained.

In January 1956 British Waterways announced that £1,396,000 was to be spent on the canal up to Berkhamsted[6]. The plans included enlarging, duplicating and mechanising the locks at Brentford, lowering lock and bridge cills up to Berkhamsted to allow ninety-ton barges to operate, and extensive bank protection and dredging.

Work started soon afterwards, and the new duplicated Thames locks were opened formally on 24 October 1962, enabling 175-ton barges to reach Brentford depot[7].

Considerable work was also carried out north of Berkhamsted. A concrete pile making plant was established at Marsworth, many miles of bank protection work was carried out, together with much dredging, and there was an extensive programme to modernise plant and equipment.

As the Board of Survey proposals had generated considerable criticism, the Government appointed an independent committee in 1956 to review the future of the waterways. Chaired by Leslie Bowes, the committee reported in July 1958, putting the waterways into three similar categories, but recommending that the main system should be put into good order with Government aid, and so maintained for at least twenty-five years.

The badge of British Transport Waterways. The same badge, with different lettering, was used by the earlier Docks and Inland Waterways Executive and the later British Waterways Board.

From 1 January 1963 the British Transport Commission was scrapped, and an independent British Waterways Board was set up to control the canals. In January 1964 the new board produced an interim report, which concluded that only that part of the Grand Union up to Uxbridge had any real commercial prospects. Whilst most of the rest still carried some commercial traffic, it incurred heavy losses. A further report followed in December 1965 which discussed the economics of each canal and the various options available, all of which involved Government funding.

The implied closure threat to a large part of the country's canal system led to a strong opposition. Following a public meeting at Watford in Hertfordshire on 16 May 1967, a Grand Union Canal Society was formed, pledged to fight to save all the canal, and particularly the Aylesbury branch, which had last been used for commercial traffic in July 1964, and was seriously threatened.

In September 1967 the Government produced a White Paper outlining a network of canals that were to make up a system of amenity cruising waterways. It was formally incorporated into the 1968 Transport Act and included most of the Grand Union system that was still open.

By then parts of the canal had already gone. The Wendover branch beyond Tringford remained closed, whilst in 1961 the Old Stratford and Buckingham branches were legally abandoned, as both had been unused for years. Their last traffic had been in 1935, but then only to Deanshanger, three miles up from Cosgrove. Part of the Erewash Canal, north of Ilkeston, was also abandoned in 1961.

A notable omission from the cruising network was the branch to Slough, last used for commercial traffic in March 1960. Road building plans in the town threatened the branch's survival, and the Grand Union Canal Society mounted a vigorous campaign to save it. The branch remained navigable with difficulty, and eventually the closure threat was lifted. Dredging took place in 1972, and in 1983 the branch was officially included in the cruising network[8].

The canal basin at Market Harborough in 1958, when commercial traffic had finished and pleasure craft had yet to appear. Early in 1964 Harborough Marine Ltd started a hire cruiser base and a boatbuilding business here. On the left Trenery's timber wharf used to be supplied with timber imported through the Humber ports.
Lewis Edwards

Elsewhere an Erewash Canal Preservation and Development Association was formed, and succeeded in reopening the abandoned part of that canal in May 1973. The short Welford branch on the Leicester line had last been used by commercial craft in 1946, had been unnavigable since 1957, but was important as a feeder from Naseby, Sulby and Welford reservoirs. The Old Union Canals Society reopened it in May 1969, whilst at Warwick the Saltisford Canal Trust reopened most of the short Saltisford branch in July 1985.

These developments were due mainly to the large growth in pleasure craft on the canal. In turn this had been influenced by the campaigning work of the Inland Waterways Association, which had

A year after the re-opening of the Welford branch in 1969, a few pleasure boats have started to appear at Welford Wharf.
Lewis Edwards

been founded in 1946, and succeeded in transforming public opinion into safeguarding the canal system.

As early as 1950 British Waterways had adapted two of its boats to provide occasional trips in the Uxbridge and Watford areas. On 5 May 1951 a regular boat service was inaugurated by John James on the canal at Paddington[9], as part of the Festival of Britain celebrations. Using the narrow boat *Jason,* passengers were taken along the Regent's Canal past the London Zoo to Hampstead Road, and back. The venture proved most successful.

Another notable trip boat, *Arcturus,* run by Bryan Nicoll, began operating on the canal at Berkhamsted in 1965, but since 1966 has been based at Cassio Bridge, Watford. Yet another, *Linda,* run by Charles and Anne Crossley, started trips at Foxton in 1956, but by 1966 had been moved to a more permanent base at Cosgrove.

In 1950 New Way Holidays started using two boats as a floating hotel. In 1953 the Inland Waterway Cruising Company began operating another pair of hotel boats from Braunston, and others followed. A hire boat business, the Floating Enterprise, was established at Uxbridge in 1955[10]. Later it moved to Watford, and it was the forerunner of many others. In 1956 the Wyvern Shipping Co. Ltd. of Leighton Buzzard, which had started as a commercial carrier in 1954, was faced with a decline in traffic, and converted some of its boats to hire craft.

Another hire boat business started at Braunston in 1957 when Blue Line Cruisers was formed. Concoform Marine moved from the Thames to set up at Weedon in 1969, and Bridgewater Boats was established at Berkhamsted in 1971. There are now yards that hire out craft to cover all the Grand Union system.

There has also been a large growth in private cruising. On 1 April 1959 the Aylesbury Boat Company established a marina at Aylesbury[11], which helped to save the branch from the closure threats. Similarly a boatyard at Iver on the Slough branch, and the moorings it provided, were instrumental in helping save that branch. Since then marinas have been established throughout the canal, such as at Whilton, at the bottom of Long Buckby locks in 1971, at Stowe Hill in 1978, at Milton Keynes in 1988, at Cowroast in 1980, and at Harefield in disused gravel workings in 1986.

A towpath bridge carrying the Slough branch path over the main line.

There was a major development at Braunston, where the third and largest engine reservoir was leased to Blue Line Cruisers in 1960. It was then connected to the Braunston branch, to enable it to be used for moorings. Soon after, the second reservoir, already linked to the first, was connected to the main canal, and a footbridge, salvaged from the closed Buckingham branch, was erected across the breached towpath in February 1962. Both were then also used for moorings, and were later leased to Blue Line. In 1969 the second and third reservoirs were connected, and a new bridge was provided to carry a public footpath, all three reservoirs now forming part of the Braunston marina, and providing berths for 500 craft.

Private boating clubs have also developed. The pioneer was the

West London Motor Cruising Club, established at Alperton in 1938. At Cheddington the Dunstable & District Boat Club, founded in 1964, has its own moorings and facilities. The Mid Warwickshire Yacht Club, founded at Hatton in 1960, and the Soar Boating Club at Kegworth in 1953, are other examples.

Since 1929 there have been several important developments over the all-important provision of adequate water supplies. Impetus was given by a series of dry years, culminating in serious droughts in 1934 and 1944. In August 1934 the position was so serious that the Leicester line was closed to conserve water, and temporary pumps were installed at Foxton to lift water from Saddington reservoir up to the summit, from where it could be run down the Watford locks to supply the main line.

At the same time steam traction engines were brought in at Long Buckby to drive pumps at each lock[12]. Soon afterwards five permanent electric pumps were installed, with one lifting water round three locks. In 1992 the individual pumps were replaced by a single unit, returning the water from the bottom to the top of the flight.

During the drought temporary pumps were also used at Stoke Bruerne. At the same time the Buckingham branch was cleaned out to allow water from the Great Ouse at Maids Moreton to feed into the canal at Cosgrove. After the Second World War, by which time the branch was impassable, emergency supplies were sometimes drawn from the river Tove at Stoke Bruerne, or from the Great Ouse at Wolverton.

In 1939 a new pumping station was built at the bottom of Stoke Bruerne locks, and a 14in diameter main was laid from it to discharge above the top lock. A diesel-driven pump was installed in 1940; it was electrified in 1948. At Braunston the steam engine was replaced by a Ruston diesel in 1940, and the pump was electrified in 1958.

On the Tring summit a pump powered by a petrol engine was installed at Weston Turville in 1934 to enable water to be fed into the Wendover branch through a temporary pipeline across the fields. This reservoir had not been used to supply the summit for many years, but in August 1944 a permanent electric pump was installed[13].

There was another crisis in 1944. As early as December 1943 the Aylesbury branch was closed to conserve water supplies, with goods having to be transhipped at Marsworth to and from lorries. It was not reopened until April 1947, and the Leicester line was closed at the same time.

To increase supplies on the Tring summit a 66ft borehole was made from the bottom of the 78ft well at Cowroast, and a 170ft borehole was sunk at Northchurch lock. The shortages were already so bad in February 1943 that a temporary pump was installed at Northchurch to deliver water into the canal. Electric pumps were provided at Dudswell and Cowroast in February 1944 to lift this water to the summit. A new engine house and permanent electric plant was commissioned at Northchurch in July 1945, and new electric plant was installed at Cowroast in November[14]. In May 1944 emergency supplies had to be taken from the river Ousel below Grove lock 28, from where the Northern Engines lifted the water to the reservoirs. These engines were progressively electrified between 1946 and 1964.

At Tringford two new electric pumps were installed in May 1945 to pump direct from the Wendover pipeline. Their speed, and hence their output, was controlled automatically by the quantity of water in the pipeline. Then in October 1960 the diesel-electric pump at one well was converted to mains electricity, and this was followed by the other well in March 1961.

There were also developments at the two main tunnels, as on 30 September 1936 the steam tunnel tug services were discontinued[15]. A few years earlier 175 horse-drawn craft had passed through every

In 1958 British Waterways inaugurated a water bus service from Little Venice, Paddington, to the London Zoo in Regent's Park. Here the first boat, *Water Buck*, is waiting at Little Venice for passengers. *Lewis Edwards.*

Opposite page: Between Soulbury and Stoke Hammond, north of Leighton Buzzard, is this unusual turnover bridge; because of the way the bridge approaches are set out there is no need to unhitch the towline as the horse crosses the bridge. The boat is Samuel Barlow's *Dragonfly*, built by Nurser at Braunston.

week, but the number had declined to a mere dozen, as by then many were self-propelled.

This, though, had no effect on the repairs which continued to pose a serious problem at Blisworth. Following several stoppages the tunnel was closed early in 1980 for major repairs, but for a few months limited access was allowed for boats to pass through. It was closed completely in the autumn as parts were in a dangerous condition, following considerable deterioration to the brickwork. Eventually in December 1982 a £4.3 million contract was awarded to John Mowlem Ltd. to completely rebuild the central 1,000yd section of the tunnel, using precast concrete segments. It was reopened on 22 August 1984.

Even Braunston tunnel gave problems. It was closed in October 1978 for urgent brickwork repairs, and not reopened until March 1980. Then both Crick and Saddington tunnels on the Leicester line were closed early in 1979, due to serious deterioration in their linings. Crick was reopened in June, but repairs to Saddington were not completed until November 1980.

Another development was the opening on 4 May 1963 of a new canal museum in the former mill at Stoke Bruerne. A boat weighing machine, salvaged from South Wales, was erected in the duplicate top lock nearby to form part of the museum exhibit. The museum has expanded since, with a new shop being opened in 1970, and is now a popular venue.

Today, with the growth of pleasure traffic, the Grand Union is as busy as it has ever been and it provides a major amenity and important leisure facilities. In this changed role it can only have an assured long-term future as a national asset.

NOTES TO CHAPTER TEN

1 21 & 22 Geo V, c 107—31 July 1931.
2 21 & 22 Geo V, c 90—31 July 1931.
3 Authorised retrospectively in 6 & 7 Geo VI, c 5—6 July 1943.
4 Minutes of evidence by John Miller, GUC chairman, to the 1943 Bill.
5 *The Times,* 9 May 1945.
6 *Waterways* (British Waterways Staff Magazine), November/December 1957.
7 *Modern Transport,* 10 November 1962.
8 Royal Assent given on 8 February 1983. The Act included the Erewash and several other canals.
9 *The Bulletin* (Journal of the Inland Waterways Association), May 1951.
10 Ibid, Summer 1955.
11 Ibid, July 1959.
12 Information from the late C.N. Hadlow, former District Inspector.
13 Richardson, Alan. Water Supplies to Tring Summit (Part 2). *Journal of the Railway & Canal Historical Society,* July 1969.
14 Ibid.
15 *The Times,* 29 September 1936.

Overleaf: Working up through Startopsend lock, lowest of the flight of six leading up to the Tring summit, are the Fellows Morton & Clayton steamer *Phoenix* of 1893 and her butty, laden with Birmingham ware for export.

DRAGONFLY.

BOXMO

The Carriers 11

AS MIGHT BE expected on a major trunk waterway, the larger firms of canal carriers always dominated the long-distance trade on the Grand Junction.

One of the earliest was Pickfords, who had an extensive trade between Manchester and London, and started using the canals for part of the journey in 1786[1]. Soon after the Grand Junction was opened to Blisworth in 1796, the firm established a wharf there. Goods were brought by boat from the north and transhipment then took place to road vehicles for the remainder of the journey to London. Once the railroad was opened over Blisworth hill, the firm established a fleet of boats on the southern section of the canal to handle its traffic down to London.

A wide variety of general goods was carried and interconnecting services throughout the country were provided. For instance, on the day the Grand Union Canal was opened in 1814 two Pickfords fly boats made their first direct run to Leicester, and the firm then leased a wharf in the city.

Soon after the Paddington branch was opened in 1801 a large wharf was leased from the Grand Junction in Paddington Basin. For the next thirty years the firm was easily the most important single carrier on the canal, with its monthly toll account often exceeding £4,000. In 1820 the firm had over eighty craft in service, and by 1838 the number had risen to 138.

Normally Pickfords enjoyed three months' credit, but they did not always stick to these terms. In September 1816 they owed £4,473 for May's tonnage and rent, and the Grand Junction got no response to a letter. It then issued an ultimatum saying that unless £2,000 was paid by 2 October Pickfords' credit would be stopped at Paddington. These threats brought their Paddington agent to a meeting at the end of October, by which time the July tonnages of £2,298 were nearly due. He pleaded against the suspension, and promised prompter payment in future. His diplomacy succeeded, but this was not the last time that Pickfords' debt caused the company concern.

In December 1820 Pickfords moved from Paddington to the City Road Basin on the Regent's Canal, as this had better access to the City of London. With the coming of the railways, Pickfords began withdrawing some of their canal services, and transferring to rail transport. The process was accelerated by the formation of the L&NWR, which effectively dictated Pickfords' withdrawal from canal carrying. The last canal traffic, to and from Manchester, came to an end on 31 December 1847.

Faced with the impending loss of this trade, the Grand Junction

Opposite page: Loading a lock gate into a maintenance boat at the Grand Junction workshops at Bulbourne, near Tring. The gate, being lowered carefully by a gantry crane, will have to be stowed on its side; the balance beam will be fitted after installation.

set up its own Carrying Establishment with a view to retaining the traffic for the canal, and building up a fleet of boats. A series of carriers was acquired, the first being Wheatcroft & Company of Cromford at the end of November 1847. Its boats were soon joined by many from Pickfords, with the Grand Junction taking over the important Manchester trade from the beginning of 1848. This averaged seventy-five tons daily from Manchester, and thirty-five tons daily from Paddington, with a considerable amount of intermediate traffic.

Three businesses were taken over in October 1848: Soresbys of Cavendish Bridge near Derby; Worster & Company with bases at the City Road Basin, Long Buckby, Northampton and Market Harborough; and John Whitehouse & Sons of Dudley. Others followed including Crowley & Company of Wolverhampton, Corbett & Sons of Stourbridge, Job Bissell of Tipton, and William Ebbern who traded between London and the Potteries.

Many of the former owners continued as agents for the company, which built up a chain of depots such as at Derby, Burton, Lichfield, Tipton and Warwick. In September 1848 a wharf was leased in Leicester, and in December one in Birmingham. An agent was appointed in the Potteries in March 1849, and in the following month land was bought for a wharf at Wolverhampton. In London the business was first based at Paddington, but several of the firms purchased had operated from the City Road Basin, and in April 1849 the company moved its carrying operations there.

In April 1855, following concern about losses, carrying was scaled down, with the traffic in heavy iron castings from

A group of FMC boats at the junction with the Oxford Canal at Braunston in 1923. *Ike Argent collection*

Birmingham to London being transferred to Price & Son and to The Trustees of the late James Fellows. Early in 1856 Mellor Colsell & Company took over the cheese and pottery trade to London, and purchased eighteen boats for it. In 1865 this firm took over the Manchester traffic as well.

In 1859 the company fitted up one of its boats, *Pioneer,* with a steam engine and screw propeller at its Wenlock Basin yard on the Regent's Canal. She proved satisfactory, and in December 1859 two new tugs were ordered together with nine carrying boats, all equipped with steam engines. The tugs were used on the Long Level between Paddington and Cowley, but the other boats joined the ordinary carrying fleet. To pay for them £4,000 was borrowed from the company's bankers.

There was a catastrophe early on the morning of 2 October 1874 when six of the company's boats were on the Regent's Canal being towed by the steamer *Ready* on their way from City Road Basin to Nottingham. One of the towed boats, *Tilbury,* was carrying five tons of gunpowder and several barrels of petroleum. As the procession was passing under Macclesfield bridge *Tilbury* blew up, killing her crew of three instantly. The *Limehouse* immediately behind her was sunk, and the *Dee* in front badly damaged and her steerer injured.

Macclesfield bridge was destroyed and many houses nearby were severely damaged. The accident was caused by a lamp, or the cabin fire, igniting petroleum vapour[2]. The canal was completely blocked and not re-opened until the afternoon of 6 October, despite 100 men working day and night.

At first the company tried to deny liability for the damage, but a court case decided against it. Eventually over 800 claims were received, and nearly £80,000 paid out in compensation. The company's bankers came to the rescue with a £40,000 loan, which was not repaid until the end of 1877.

As higher costs had increased the carrying losses in 1875, this disaster made the company decide to give up carrying from 1 July 1876. New and existing carriers now stepped in to take over its trade, and many of its boats.

Prominent among the new carriers was the London & Staffordshire Carrying Company founded by Joshua Fellows, the senior partner in The Trustees of the late James Fellows, and Hugh Hughes, the former manager of the Carrying Establishment. The new company leased, and later purchased, several of the Grand Junction's boats and steamers. Operating from Brentford and the City Road Basin, the company carried mainly to Birmingham, Wolverhampton and Stourbridge.

Another new concern was the London & Midland Counties Carrying Company, set up by several former Grand Junction agents to retain the trade to the East Midlands. It also leased several of the boats, and had depots at City Road Basin, Leicester, Nottingham, Sandiacre and Derby.

At the same time Fellows Morton & Co was created; a merger of the late James Fellows' business with that of Frederick Morton of Wolverhampton. Fellows had set up as a carrier at West Bromwich in 1837, and began trading down to London in 1839, later moving to a base at Tipton. He died in 1854, but his trustees not only continued but developed his business with several branches being established, including one at Brentford. Frederick Morton had become involved in carrying in 1858, had acted as agent for the Bridgewater Canal at Wolverhampton, and later for the Fellows' trustees.

Fellows Morton & Company went on to acquire the London & Staffordshire Carrying Company in 1879 and the London & Midland Counties Carrying Company in 1887, along with several other firms, making it the dominant long-distance carrier on the Grand Junction.

Its position was further strengthened in 1889 when the general carrying side of William Clayton of Saltley's business was taken in. A new company, Fellows, Morton & Clayton Ltd., was incorporated to control the combined concern[3]. Other acquisitions followed, including George Hurst & Son of Northampton and Leighton Buzzard, the London & Birmingham Canal Carrying Co. Ltd. and Samuel Phipkin of Hillingdon.

FMC operated a fleet of steamers on its main traffics between London and the Midlands. On the Grand Junction depots were maintained at the City Road Basin and Brentford, and in 1896 a boatyard and coal wharf were acquired at Uxbridge. In 1912 it

The King, one of FMC's famous fleet of steamers, under way at Long Buckby in 1910. *National Waterways Museum*

pioneered the introduction of motor boats powered by semi-diesels, and by the mid-1930s had ceased using horses.

The company handled a wide variety of traffics including many foodstuffs for the Midlands, with sugar in sacks, chests of tea, canned goods, sultanas and spices from abroad being typical cargoes. Newsprint, chemicals, timber and grain were others. Manufactured goods were carried down to London, mainly for export, and another southbound traffic was beer, delivered to Park Royal on the Paddington branch.

FMC's trade was badly affected during the First World War and it incurred a loss. The Second World War caused similar disruption resulting in another loss. The company went into voluntary liquidation in November 1948, and most of its assets were acquired by the British Transport Commission.

As part of its development programme, the Grand Union Canal Company had acquired in 1929 an experimental pair of narrow boats built at Uxbridge. A small carrying company—Associated Canal Carriers Ltd. of Northampton—was taken over in 1930 and a modest start made to enlarge its fleet[4]. In 1934 this carrying company was renamed the Grand Union Canal Carrying Co. Ltd., and an ambitious scheme was launched to increase the fleet to 100 pairs. A further eighty-six pairs were ordered in 1936. A depot was set up at Bulls Bridge in Southall in 1936, opposite the entrance to the Paddington branch.

There was also a drive to find new traffics, and large amounts of steel were obtained, imported from the Continent through Regent's Canal Dock for the Midlands. Considerable volumes of chemicals, non ferrous metals, grain, timber and straw board were also secured, mostly from competing road and rail transport[5]. Some traffics were taken from other canal carriers. In 1938 a contract was won to supply coal to the John Dickinson mills at Apsley, Nash and Croxley, which previously had been handled by a consortium of owner boatmen.

Despite its large investment the Grand Union could not make its carrying subsidiary profitable, due to the difficulty in finding sufficient crews to man the boats. As a result many of the craft had to be laid up, and the active fleet never exceeded 119 pairs. In 1939 the company started to dispose of spare boats, and by the end of December 1947 the fleet had been reduced to 126 pairs.

On nationalisation these boats formed the basis for British Waterways South Eastern Division fleet. From 1 January 1949 they were joined by many of the boats operated by FMC, making British Waterways by far the largest carrier. It went on to invest considerable sums in re-equipping and modernising its fleet, and even investing in new boats[6]. Despite this the decline continued. By 1956 the fleet had reduced to 100 pairs, by 1959 to eighty, and by the end of 1962 to forty-five pairs.

Canal carrying was now coming under increasing competition, with a steady loss of contracts, particularly in the main southbound traffic of coal from the Warwickshire collieries. Businesses were turning to other forms of transport, to other fuels such as oil, or to new locations not served by water transport.

The loss of the contract to carry coal to Dickinson's Nash mill was typical. The last delivery was in July 1962, the factory then switched to oil firing. The Apsley mill contract was lost soon afterwards, as the supply was reallocated to a colliery not served by canal. Home Park, Dickinson's other mill in the area, was served by a private carrier and deliveries ceased in the autumn of 1960.

By now the nationalised fleet was incurring substantial losses, and following a hard winter, when the boats were frozen in for several months, British Waterways gave up almost all carrying early in 1963.

The Willow Wren Canal Carrying Company was formed at the end of December 1953 to try to reverse the decline in canal carrying[7]. With generous backers, a small fleet of boats was built up and several traffics were regained for the canal. Typical was coal to the Nestle factory at Hayes, roadstone to West Drayton, timber from Brentford to Aylesbury, Bletchley and Olton, and coal to a factory at Slough.

Despite initial successes Willow Wren was unable to trade profitably. It ceased carrying early in 1963, but most of its assets were passed over to a new concern, Willow Wren Canal Transport Services Ltd., which had been created in an attempt to retain the remaining traffics. It also took over the traffics from British Waterways, and leased many of its boats. Despite this the decline continued, with the coal contract to Dickinson's Croxley mill finishing early in 1967, owing to the closure of the colliery, although a part was regained later for a short time. The last main traffic, grain from Brentford to Wellingborough on the river Nene, was lost to the railways in 1969. The company then went into receivership, and ceased trading after making the final delivery to Croxley in September 1970.

One traffic retained by British Waterways in 1963 was the transport of barrels of lime juice from Brentford to Hemel Hempstead. In 1973 the contract passed to an independent carrier, but deliveries finally ended in October 1981.

There were many other long-distance carriers on the canal. One was Lawrence Faulkner of Leighton Buzzard, who set up in 1899 and handled much of the silica sand traffic that was despatched from the area[8]. This went to the glass making industry at Spon Lane near Birmingham, and to Paddington and Brentford for the building industry. He also brought in coal, grain, animal feeds and timber to Leighton. Faulkner died in 1950, and the remaining boats were sold off in 1951.

Another carrier was Arthur Harvey-Taylor, who took over the old-established business of John Landon & Company in 1923. Harvey-Taylor handled most of the traffic to and from Aylesbury[9]. Coal was delivered to the town's power station, Nestle's dairy, the Aylesbury Steam Laundry, and domestic consumers. Other traffics

The FMC steamer *Sultan* and its butty *Ash* at Syon Park, just north of Brentford, in September 1903. *William Knibbs*

were grain, straw board and timber from Brentford, and sand from Leighton Buzzard. He also delivered large quantities of grain to Mead's flour mill at Tring on the Wendover branch. After the Second World War many of the traffics were lost, and the last two pairs were sold in 1955.

Some businesses specialised in a particular traffic. The Samuel Barlow Coal Co. Ltd. was a coal dealer, and developed contracts to supply many factories and wharves[10]. Typical contracts on the Grand Union were to the Aylesbury Dairy, the British Paper Company's mill at Frogmore near Hemel Hempstead, the Ovaltine works at Kings Langley, the British Portland Cement Company's works at Harefield, the Uxbridge Electric Light Works, the International Tea Company's jam factory at Southall, and many others.

The remnants of the Barlow fleet were taken over by Blue Line Canal Carriers Ltd. in 1962, which continued carrying until the closure of the jam factory at Southall brought the final contract to an end in October 1970.

Another fleet which only handled coal was that operated by A. Wander Ltd. of Kings Langley. This company had established its Ovaltine factory beside the Grand Junction in 1913, and was supplied with coal by the canal[11]. Initially Barlows handled most of the deliveries, but in 1926 the company started introducing its own boats, and eventually it had nearly a dozen pairs. Canal deliveries ceased in April 1959, with the traffic passing to road transport, but soon afterwards the factory switched to oil firing.

The heaviest volumes of traffic were carried in barges at the southern end of the canal, albeit usually over short distances. The largest operator was the Thames Steam Tug & Lighterage Co. Ltd., which was incorporated in 1856. In 1960 it had over 600 barges, of which nearly half could use the canal[12]. They loaded from, or unloaded to, ships in the London docks, and the company ran a fleet of tugs to tow them between the docks and Brentford. Some discharged their cargoes into the warehouses at Brentford depot, or into narrow boats. Typical traffics handled in this way were tomato puree from Italy, lime juice from the West Indies, wheat from Canada, timber from the Baltic, with sugar, paper, cotton, zinc and much more.

In many cases the barges continued on up the canal, being towed by horses, or later by small towpath tractors. Oil from refineries at Shellhaven and Purfleet was delivered to the large Castrol depot at Hayes, timber went to several wharves up to Uxbridge, milk powder to Glaxo at Greenford, and sleepers to the railway depot beside the Paddington branch at Southall.

Several barge owners were based in the Brentford area, one of the best-known being the Bucks. By the end of the 1880s both Alfred and John Buck were established at Norwood top lock, and soon afterwards William set up at Brentford[13]. From 1925 the family interests were consolidated into Alfred Buck & Son Ltd. at Brentford. The company was the principal barge operator up to Berkhamsted until well after the Second World War.

George Charles Collier set up at Brentford in the mid 1880s[14], and eventually had over twenty barges. Following his death in 1908, most passed to his son, George Henry. The business was based at Brent Meadow Wharf, and was incorporated into G. H. Collier Ltd. in 1926. Colliers were the main contractors to serve Lyons dock.

James Clements and George Knowling traded together as Clements Knowling & Co from Goat Wharf at Brentford. Eventually the business was merged with that of Tough & Henderson Ltd to form Clements Tough Ltd., a company associated with the John Dickinson group. Its barges were used on the run up to Croxley mills with such cargoes as esparto grass.

In the mid-1890s Emanuel Smith began trading from Brentford, and he built up an extensive fleet. In 1916 he had four steam tugs operating on the Thames, and was actively involved in the lighterage trade.

He also operated a fleet of wide boats. These were built on similar lines to a narrow boat, but were between 9ft and 13ft in width, and had rounded sides to reduce water resistance. In theory wide boats were able to navigate the entire main line, but in practice their use was normally confined to the southern section. Smith also operated a fleet of narrow boats and was one of the pioneers of motor boats, of which the first, *Premier,* was built at Rickmansworth in May 1912. The business was based at Ham Wharf on the semi-tidal part of the canal, above Thames lock. It was incorporated to become Emanuel Smith & Sons Ltd. in 1923, but then ran into financial problems and went into liquidation in 1930.

Another Brentford-based company was Thomas Clayton (Oldbury) Ltd., which specialised in the transport of bulk liquids and operated a fleet of barges, wide boats and narrow boats, all decked over as tankers[15]. Wide boats collected tar from gasworks as far north as Long Buckby for delivery to Brentford for distillation. Other works served were Newport Pagnell—from Great Linford basin—Fenny Stratford, Berkhamsted, Hemel Hempstead, Watford, Rickmansworth, and Pinner—from Northolt on the Paddington branch.

Initially narrow boats were used for those gasworks the wide boats could not reach, such as Northampton, Aylesbury, and possibly Wendover, but later they took over many of the longer-distance runs. They also made deliveries to distilleries in the Midlands, especially from the Northampton works.

Large quantities of tar were collected from the Uxbridge and Southall gasworks. The barges used for this remained horsedrawn until the traffic finished in 1960; the company was probably the last to use horses on the canal. Other traffics included fuel oil, with regular deliveries being made to both Southall and Uxbridge gasworks.

Another series of businesses was based at Paddington and most of these operated fleets of wide boats. The largest, Odell & Co. Ltd., delivered building materials into Paddington basin, and transported many of the waste products of the capital away. In 1897 Thomas Clayton, who was already involved at Brentford, purchased the business from Henry Odell. The company, renamed Thomas Clayton (Paddington) Ltd. in 1910, had a fleet of nearly seventy craft, of which over half were wide boats[16]. The company had held the contract to handle Paddington's rubbish since the end of the

A pair of narrow boats belonging to Emanuel Smith of Brentford tied up at Long Buckby locks. Between 1912 and 1914 W. H. Walker & Brothers at Rickmansworth built five motor boats for Smith, who was among the first to introduce powered craft of this kind into his fleet. *National Waterways Museum*

1890s, and in the 1930s this averaged 42,000 tons each year.

St Marylebone Borough Council used its fleet, set up in 1918 when the craft operated by the Western Cartage Co. Ltd.[17] were purchased, to transport rubbish from the borough to tips at Yeading and Hayes. The boats were towed in trains along the Long Level by small tugs. From 1921 some wide boats were built and eventually the council had two dozen craft in service. In August 1939 its fleet was taken over by Thomas Clayton (Paddington) Ltd., which obtained a fifteen-year contract to handle the borough's rubbish.

H. Sabey & Co. Ltd. of South Wharf, Paddington, was involved in the sand and gravel business. It constructed the side lock on the Rickmansworth branch in 1902 to enable the gravel to be extracted from Batchworth Meadows, and also had a pit at Yiewsley served by a dock. From 1911, the company developed its own fleet of wide boats. Initially these were second-hand craft, but a new boat was built in 1913 and others followed, with ten between 1930 and 1932[18]. In 1935 there were twenty-five boats in service, making it second only to Claytons. Sabeys' canal traffic dropped sharply during the Second World War and ended soon after.

In 1962 the Thames Steam Tug & Lighterage Co. Ltd. merged with General Lighterage Ltd., established in 1922, to form Thames & General Lighterage Ltd. This made it by far the largest barge operator on the canal, but later it also took over Clements Tough Ltd. and Vokins & Co. Ltd. Vokins was another old-established company, which had acquired most of the barges operated by Alfred Buck & Son Ltd. It held the important contract to deliver grain to the Quaker Oats factory situated on Maypole Dock at Southall.

The wide boat *Frays* in Stoke Bruerne top lock in 1914. This floating tanker was on her way up to Long Buckby gasworks to collect crude tar for transporting to Brentford for distillation. She was operated by Thomas Clayton (Oldbury) Ltd. *National Waterways Museum*

Following the progressive closure of London's dockland there was a rapid rundown in the lighterage trade. Eventually even the massive Thames & General company gave up, but to try to retain water-borne traffic to its Brentford depot, the British Waterways Board purchased a fleet of tugs and barges in February 1982, and set up Lee & Brentford Lighterage. After a dock strike and other difficulties, this ceased trading in November 1984, bringing all barge operations on the canal to an end.

There was a series of boatyards along the canal to build and service these craft. In 1801 there was a yard at Braunston run by a Mr Hughes, but this may have been situated on the Oxford Canal, where William Nurser started as a boat builder in 1875. Nurser's business remained in the family until 1941, when it was purchased by the Samuel Barlow Coal Co. Ltd. It was still an active yard when Blue Line, which took over from Barlows, gave up canal carrying in 1970, although by then most of its work derived from pleasure craft.

There was also a yard near Braunston bottom lock, and latterly this incorporated the gauging dock that had been built there in 1816. In 1954 Willow Wren acquired and improved the yard, but sold it to Braunston Boats Ltd. in 1968.

In 1866 William Thompson built a boat dock into his wharf at

High House bridge, Nether Heyford, but this closed down probably in the 1920s.

Not all boatbuilding was carried on alongside the canal, for at Stony Stratford Edward Hayes, who had set up as an agricultural engineer in 1840, and his son, another Edward, produced marine engines and subsequently built boats in a yard some distance from the canal. Their craft, many of which were for export, had to be taken on trailers to be launched into the Old Stratford branch. The son died in 1920 and the yard closed down in 1925[19].

Another early, and probably short-lived, yard was at Fenny Stratford where George and John Johnson were established as soon as the canal opened in 1799. At Leighton Buzzard Lawrence Faulkner starting repairing his own fleet in the 1910s, and later he took over Charity Wharf in Linslade as a small boatyard. It operated until 1956, when it closed through lack of business.

In the 1850s John Bushell had started to repair some of the boats that operated to Mead's flour mill at Tring on the Wendover branch. His son Joseph developed this into a separate business in 1875 from a yard alongside the mill. In 1912 grandsons Charles and Joseph took over, trading as Bushell Brothers, and they remained in business until their retirement in 1952[20]. A wide variety of craft was built and repaired; one customer was Arthur Harvey-Taylor of Aylesbury.

A yard had been opened at Berkhamsted by Peacock & Willetts soon after the canal reached there in 1798. By 1826 it was operated by John Hatton, and William Edmund Costin took it over from

The barge *Tiny*, built for Charles Beckett of Kingston at Bushell's boatyard on the Wendover branch, awaits her launching in 1918. *National Waterways Museum*

Opposite page, top: An early view of Walkers' boatyard at Rickmansworth showing FMC's *Lion,* which may have paused for repairs. *W. H. Walker & Brothers Ltd*

Opposite page, bottom: Costin's boatyard at Berkhamsted, with the staff posing on the steam tug *Kestrel,* which had just been launched for Tough & Henderson, who operated barges at the southern end of the canal. *W. H. Walker & Brothers Ltd*

Below: A general view of Walkers' boatyard in 1913 with a wide boat on the dock and other craft, including the wide boat *Alice May* which had just been launched for Joseph Heather of West Drayton, moored alongside. *W. H. Walker & Brothers Ltd*

John Joseph Hatton in about 1882. Costin's business, incorporated to become W. E. Costin Ltd. in 1899, built and repaired barges, wide boats and narrow boats for many customers including Emanuel Smith and Lawrence Faulkner, until it closed in 1910.

In 1905 William Henry Walker, who had served an apprenticeship as a boatbuilder with Bushell Brothers, set up a boatyard at Frogmoor Wharf, Rickmansworth[21]. In 1922 the business was incorporated to become W. H. Walker & Brothers Ltd. The yard built over 100 narrow boats for the Grand Union Canal Carrying Company between 1934 and 1938. Customers included Thomas Clayton (Oldbury) Ltd., Cadburys of Bournville, A. Wander Ltd. and many others. The company also traded as coal and builders' merchants, receiving many of its supplies by canal. The last new narrow boat was built in 1952, but repair work continued until 1964.

Another important yard was at Uxbridge, and was acquired by FMC from Edward Morgan in 1896. Many of the company's boats operating on the canal were built or repaired there, the last new boat being built in 1947.

At West Drayton a boatyard was operated by William Garrett and Walter Houghton in the 1910s. It was based at Cooper's Dock and was previously run by Edward Dutton. The Grand Union Canal Carrying Company established a boatyard at Bulls Bridge in 1936. Following extension of the dry dock facilities in 1956, the yard took over all the repair work on the British Waterways fleet, whereas

FELLOWS, MORTON
& CLAYTON LTD
No.38.

previously much had been contracted out. It remained in business after narrow boat carrying ended to service the fleet of maintenance craft.

Brentford was a barge-building centre. For many years Edward Woods was established at Staffordshire Wharf, near Thames Lock. The yard was purchased in 1953 by Vokins & Co. Ltd., who changed the name to Thames Lock Wharf. Whilst dealing mainly with barges, Woods built several narrow boats, and Vokins repaired others for British Waterways. Nearby the barge operators E. C. Jones & Son established themselves at Thames lock in 1920. They became involved in barge repair work and began building new craft after the Second World War. They introduced the Bantam pusher tug in 1951 and built sixteen steel narrow boats for British Waterways between 1959 and 1961. The Thames Steam Tug & Lighterage Co. Ltd. also had a barge yard at Brentford.

There were many smaller and short-lived yards. Between them they provided the craft without which the carriers could not have handled the traffic on the canal. Some of the yards remain in business today serving the not dissimilar needs of the pleasure boat industry.

NOTES TO CHAPTER ELEVEN

1 Turnbull, Gerald. *Traffic and Transport. An Economic History of Pickfords,* 1979.
2 The Regent's Canal Explosion, London 1874, in the *Fire Protection Association Journal,* April 1957.
3 Faulkner, Alan. *F.M.C., A short history of Fellows, Morton & Clayton Ltd.,* 1975.
4 Faulkner, Alan. *The George & The Mary, A brief history of the Grand Union Canal Carrying Co Ltd.,* 1973.
5 *The Bulletin* (Journal of the Inland Waterways Association), May 1953.
6 Wilson, Robert. *Too Many Boats,* 1980.
7 Faulkner, Alan. *Willow Wren,* 1986.
8 Faulkner, Alan. Faulkners of Leighton Buzzard, in *Waterways World,* November 1988.
9 Faulkner, Alan. Harvey-Taylor, in *Waterways World,* January 1985.
10 Faulkner, Alan. *Barlows,* 1986.
11 Faulkner, Alan. The Ovaltine Boats, in *Waterways World,* March 1986.
12 *Waterways Magazine,* March 1961.
13 Brentford and Berkhamsted Canal Boat registers. (First entries 1879)
14 Brentford Canal Boat registers. (First entries 1886)
15 Faulkner, Alan. *Claytons of Oldbury,* 1978.
16 Grand Junction Canal gauging registers.
17 Paddington Canal Boat registers.
18 Paddington Canal Boat and Grand Junction Canal gauging registers.
19 *Wolverton Express,* 13 January 1961.
20 A browse through a boatbuilding family's album, in *Waterways World,* December 1989.
21 Walker, Anthony. *Walkers' of Ricky,* 1991.

Appendix 1

The Newport Pagnell Canal

THE GRAND JUNCTION had rejected the idea of a branch canal to Newport Pagnell as early as 1793, probably because the main canal was to pass fairly close to the town in any case. The people in Newport remained anxious to have a branch and approached the company again in 1802, but it still refused to act. Its funds were very short at the time and there were water supply problems in the Fenny Stratford to Wolverton pound, which would have to feed the branch.

Feelings in Newport remained strong, and in January 1813 William Praed convened a meeting in the town[1], when the merits of constructing either a branch canal or a railway from the main canal at Great Linford were discussed. A committee was set up to pursue the railway proposal, but the idea was dropped soon afterwards. A final request was then made to the Grand Junction to build a branch canal, but when this was turned down yet again the townsfolk decided to try and make the connection themselves[2].

A public meeting was held on 20 August 1813, with the local Member of Parliament in the chair[3], at which it was decided to apply for an Act to construct the canal. A subscription list was opened and £7,825 was offered immediately. George Cooch, a local lawyer, was appointed clerk and solicitor; he was destined to act for the concern throughout its existence.

At first the promoters had ambitious plans to extend the canal to Olney, or Bedford, so that the new canal would became the first part of a link between the Grand Junction and the Great Ouse, thus forming a major east-west waterway route between the Midlands and East Anglia.

The survey was carried out by Benjamin Bevan, who estimated £12,650 would be needed for the first part of the canal to Newport. His plans were approved at a meeting on 9 November, when a $2\frac{1}{2}$% call was made to meet the expenses.

There was little opposition, although the Grand Junction, which had agreed to supply water free of charge, imposed various conditions to protect its own waterway in times of shortage. The Act[4] received the Royal Assent in June 1814, and named sixty-eight promoters, who were incorporated as "The Company of Proprietors of the Newport Pagnell Canal" with power to raise £13,000 in shares of £100, £50 or £25, with a further £7,000 if necessary. Tonnage rates were fixed at a maximum of $2\frac{1}{2}$p per ton for manure, $7\frac{1}{2}$p for coal and coke, and $12\frac{1}{2}$p for all other goods.

The company advertised for tenders for carrying out the work at the end of November, specifying the canal was to be built to the same narrow dimensions and standards as the Northampton branch then under construction[5]. A contract was awarded in December, and work must have started early in 1815. The canal was to be one and a quarter miles long, with seven locks. In May 1816 the Grand Junction gave permission for the connection to be made with its waterway, and the work was completed at the end of that year.

The canal was expected to be finished on 2 December 1816[6], but the opening had to be deferred until January 1817[7], possibly because of frosty weather. A wharf at Newport was opened at the same time.

The seal of the Newport Pagnell Canal Company.

Trade developed rapidly. Coal was always the most important traffic, coming mainly from collieries in Derbyshire and Leicestershire. Other traffics were stone, manure, bricks, grain, timber, malt, oil cake and salt. For the year ending 31 May 1819 the tonnage was 9,899. It rose to a peak of 14,887 in 1845, but fell to as low as 5,205 in 1850[8]. The average was 10,160 tons per annum.

Coal tonnage was 7,212 in 1819. It rose to a peak of 11,725 in 1845, when there was a large traffic into Bedfordshire, but dropped off after the railway from Bletchley to Bedford opened in November 1846, and slumped to an all-time low of 3,669 in 1850. It then recovered to its former levels.

Toll revenue in 1819 was £434; it peaked at £728 in 1845, but dropped in 1846 when the toll on coal was reduced from 5p to 4p following pressure from the Grand Junction[9]. The average was £425.

Income from rents from the company's wharves at Great Linford and Newport produced a steady income, peaking at £280 in 1823. Thereafter it remained fairly constant until the 1850s, when it declined as tenants could not be found so readily. It averaged £185 per annum.

An initial dividend of £1.50% was paid in December 1819, and thereafter payments were generally made in June and December in each year. In 1845 the dividend was as high as £5.50%, but in other years, when maintenance costs were heavy, the shareholders received nothing. Payments averaged £2.70% a year. The canal was moderately successful, although the very large profits expected by the promoters never materialised, and the ambitious plans to extend the canal were never fulfilled.

At the end of 1845 a railway was proposed from the L&BR at Wolverton, through Newport Pagnell, to join the Bletchley to Bedford line near Ridgmont. Robert Stephenson had suggested the canal could be used for part of the route and the canal company offered to sell out for the issued share capital of £14,230[10]. A provisional agreement to buy the canal for £10,000 was reached in February 1846, and in March the canal company petitioned Parliament in support of the railway enabling Bill.

The Grand Junction opposed the transfer. Because of this opposition, and because of a general hardening of opinion in Parliament against arrangements between canal and railway companies, the scheme was withdrawn in May 1846.

At the end of 1862 a railway was proposed from Wolverton to Newport Pagnell, with possible extensions to Olney and Wellingborough. The bed of the canal between Great Linford and Newport was again to be used. The company was still keen to sell and asked £9,000 in December[11], which was agreed. A formal contract was drawn up with the proposed Newport Pagnell Railway Company and the canal shareholders approved this at a special meeting in April 1863.

Once again the Grand Junction opposed the plan, as did various

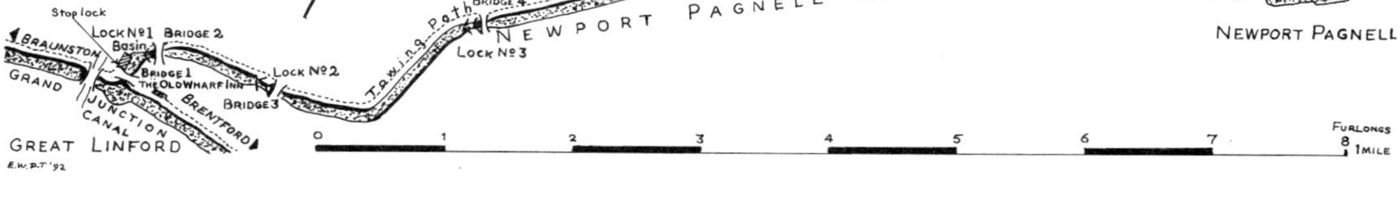

The Newport Pagnell Canal from its junction with the main line at Great Linford to the terminal basin in Newport Pagnell.

Right: Although the Newport Pagnell Canal closed in 1864 the entrance basin at Great Linford remained open. Here the Thomas Clayton wide boat *Chess* is waiting in the basin about 1910 to load a cargo of crude tar, which would have been brought by horse and cart from the Newport Pagnell gasworks. *Mansell Collection*

Below: The entrance to the Newport Pagnell Canal at Great Linford with an icebreaker on the main canal and a large team of horses on the towpath. *National Waterways Museum*

merchants who used the canal, but in June 1863 an Act[12] was passed authorising the closing of the canal and the building of the railway. The canal was closed after 8 September 1864, the railway paying over the £9,000 on the following day. A closing down sale was held at Linford wharf on 26 September, when the company sold equipment not taken over by the railway.

After settling the expenses, £8,730 remained, which was distributed to the shareholders at £61.35 for each £100 share[13]. A gratuity was paid to George Cooch for his work in winding up the company's affairs, and at the final meeting on 24 October 1864 a special vote of thanks was given him for his services of over forty-eight years. Cooch had received a small salary for his work, which was on a part-time basis. The company's only other employee was Henry Martyn, who had been its toll clerk and lock keeper for over twenty-five years. He was awarded a £10 gratuity.

The Newport Pagnell Canal used to leave the Grand Junction by the Great Linford road bridge, where there was a basin and transhipment warehouse. The seven locks were evenly spaced in their fall to Newport except for a flight of three just outside the town. At Newport there was a basin with two arms on which were several wharves and warehouses. Part of the course of the canal can still be traced at Great Linford.

The railway to Newport was opened on 2 September 1867. It was not a financial success and was eventually taken over by the L&NWR. The extension to Olney and Wellingborough was authorised by Acts of 1865 and 1867 but was never completed, although some earthworks were started along the route. The line was closed on 7 September 1964, almost exactly 100 years after the canal.

NOTES TO APPENDIX ONE

1 *Northampton Mercury,* 26 December 1812.
2 Ibid, 14 August 1813.
3 Ibid, 21 August 1813.
4 54 Geo III, c 98—17 June 1814.
5 *Northampton Mercury,* 26 November 1814.
6 Ibid, 23 November 1816.
7 *Pigot & Co's London & Provincial Directory,* 1823/4. An advertisement in the *Northampton Mercury,* 11 January 1817, describes the canal as recently opened.
8 Minutes of evidence to House of Lords Select Committee on the Newport Pagnell Railway Bill.
9 Newport Pagnell Canal minute book, 28 August 1846.
10 Ibid, 28 January 1846.
11 Ibid, 6 December 1862.
12 26 & 27 Vic, c 110—29 June 1863.
14 Newport Pagnell Canal minute book, 24 October 1864.

Appendix 2

Table of Dividends

Year:	Per cent:	Year:	Per cent:	Year:	Per cent:	Year:	Per cent:
1801	2.00 (a)	1836	12.00	1871	3.75	1906	3.00
1802	-	1837	12.00	1872	4.00	1907	3.00
1803	1.50	1838	10.00	1873	4.00	1908	3.00
1804	-	1839	10.00	1874	4.50	1909	3.00
1805	1.00	1840	8.00	1875	3.00	1910	3.00
1806	3.00	1841	7.00	1876	3.00	1911	3.00
1807	3.00	1842	7.00	1877	3.00	1912	3.00
1808	4.00	1843	7.00	1878	3.00	1913	3.00
1809	5.00	1844	7.00	1879	4.25	1914	3.50
1810	6.00	1845	7.00	1880	4.00	1915	3.00
1811	6.00	1846	6.00	1881	4.00	1916	3.00
1812	7.00	1847	5.00	1882	4.00	1917	2.50
1813	7.00	1848	5.00	1883	4.00	1918	3.50
1814	7.00	1849	5.00	1884	4.00	1919	3.75
1815	8.00	1850	4.00	1885	4.00	1920	3.75
1816	2.00 (b)	1851	3.00	1886	4.00	1921	3.75
1817	6.00	1852	3.50	1887	4.00	1922	3.75
1818	8.50	1853	4.00	1888	4.00	1923	4.00
1819	9.00	1854	3.50	1889	4.00	1924	4.00
1820	9.00	1855	2.00	1890	4.00	1925	4.00
1821	9.00	1856	2.75	1891	4.00	1926	4.00
1822	10.00	1857	4.75	1892	4.00	1927	4.00
1823	10.00	1858	3.50	1893	4.00	1928	4.00
1824	11.00 (c)	1859	3.75	1894	4.00		
1825	13.00 (c)	1860	3.50	1895	4.00		
1826	13.00 (c)	1861	3.50	1896	4.00		
1827	13.00 (c)	1862	4.00	1897	4.00		
1828	13.00	1863	4.00	1898	4.00		
1829	13.00	1864	4.00	1899	4.00		
1830	13.00	1865	4.00	1900	4.00		
1831	13.00	1866	4.50	1901	4.00		
1832	12.00	1867	4.25	1902	4.00		
1833	12.00	1868	4.00	1903	3.00		
1834	12.00	1869	4.25	1904	3.00		
1835	12.00	1870	4.25	1905	3.00		

Notes:

(a) Initial dividend declared June 1801, but not paid until February 1803.

(b) Reduction due to very heavy expenses.

(c) Including bonus payment.

TONNAGE REVENUE TABLE

Year:	£:	Year:	£:	Year:	£:	Year:	£:
1795	377	1815	147,857	1835	180,125	1855	71,679
1796	2,120	1816	127,130	1836	198,086	1856	73,886
1797	6,250	1817	145,558	1837	155,718	1857	70,521
1798	8,343	1818	169,922	1838	152,657	1858	67,365
1799	9,973	1819	157,633	1839	138,263	1859	67,070
1800	17,176	1820	151,525	1840	122,140	1860	68,896
1801	19,699	1821	159,600 (a)	1841	121,753	1861	68,873
1802	26,541	1822	153,620	1842	113,012	1862	72,702
1803	22,782	1823	169,085	1843	116,686	1863	74,007
1804	39,409	1824	178,155	1844	123,018	1864	75,638
1805	68,555	1825	187,532	1845	106,102	1865	76,830
1806	87,392	1826	178,827	1846	82,044	1866	74,894
1807	92,602	1827	189,131	1847	86,357	1867	68,422
1808	107,295	1828	181,932	1848	79,374	1868	68,531
1809	127,404	1829	181,144	1849	83,748	1869	63,996
1810	142,979	1830	176,541	1850	81,565	1870	59,741
1811	138,998	1831	168,540	1851	79.586	1871	57,905
1812	141,911	1832	167,039	1852	77,814	1872	62,537
1813	168,390	1833	170,460	1853	79,618	1873	67,601
1814	155,008	1834	174,722	1854	82,357	1874	66,082

Thereafter only incomplete records have been discovered so far.

1888 £56,455 + £28,526 (other revenue) = £84,981
1898 £64,187 + £35,888 (other revenue, rents) = £100,075
1905 £62,160 + £39,766 (rents) = £101,926

1923 £65,661 + £69,695 (rents) + £3,379 (miscellaneous) = £138,735
1924 £68,070 + £70,652 + £3,841 = £142,563
1925 £68,928 + £70,872 + £4,851 = £144,651
1926 £68,837 + £70,466 + £3,091 = £142,394
1927 £71,062 + £72,333 + £3,900 = £147,295
1928 £71,375 + £70,499 + £6,033 = £147,907

Note:
(a) Figure estimated from graph of tonnages

TONNAGE FIGURES FOR SOME OF THE MORE IMPORTANT TRAFFICS

	1904	1910	1916	1922	1928
Ashes & roughdust from Paddington	100,249	77,249	82,969	152,742	163,540
Clay & excavations from Paddington	372,263	114,688	88,275	54,135	49,356
Manure from Paddington	45,669	31,488	17,233	8,277	7,498
Slop & sweepings from Paddington	117,510	80,802	37,711	25,503	25,736
Sand & gravel to Padd'n & Brentford	83,918	50,996	68,872	97,682	108,426
Gravel & ballast south from Denham	130,779	90,009	56,024	130,877	113,936
Sand south from Leighton Buzzard	29,450	31,361	23,465	23,148	20,014
Medway ballast from Brentford	16,630	6,079	117	4,352	2,192
Roadstone from Warwickshire	10,373	9,528	4,606	7,277	6,092
Roadstone from Leicestershire	6,738	2,756	763	668	-
Lime & cement south from Harefield	14,065	19,768	19,812	23,263	30,217
Lime & cement from Stockton	5,309	502	482	137	-
Chalk from Harefield	29,258	11,938	379	5,083	4,790
Bricks north from Cowley district	11,297	15,320	2,220	430	-
Bricks south from Cowley district	104,467	28,223	7,033	2,522	797
Bricks from Gayton & Stoke Bruerne	12,117	5,097	131	19	-
Grain from Brentford	59,658	54,284	38,020	42,280	45,351
Flour from various mills	9,966	7,082	1,609	18	-
Timber from Brentford	26,720	28,736	15,088	21,587	54,217
Sleepers Brentford to GWR Hayes	20,767	21,532	15,755	14,798	15,679
Sugar from Brentford & Paddington	-	-	23,476	29,073	33,935
Milk from Aylesbury to Brentford	3,900 +	4,842	4,011	195	-
Coconut oil &c to Maypole Dock	-	-	50,528	46,329	30,574
Ale Watford to Uxbridge & Paddington	3,305	2,272	533	-	-
Ale Northampton to Braunston	3,316	1,668	939	-	-
Coal to Southall gas works	53,703	64,935	32,249	29,034	-
Coal to Uxbridge gas works	5,704	8,760	10,414	16,251	18,175
Coal Mitre Tip to Kensal Green works	19,574	30,137	-	66,813	96,352
Coal Regent's Canal to Kensal works	84,803	84,841	125,259	103,238	100,343
Oil to Southall gas works	5,819	6,695	7,667	4,936	10,744
Oil to Kensal Green gas works	5,474	4,920	3,080	7,828	8,150
Paper from Dickinson's mills	6,286	6,677	9,350	6,957	8,045
Materials from Paddington to mills	4,801	6,695	5,064	4,908	4,592

To mollify Lord Ebury, who lived at nearby Moor Park, John Dickinson had to provide his new mill at Croxley, just south of Watford, with an Egyptian-style front in 1830. *National Waterways Museum*

Materials from Brentford to mills	10,479	15,628	12,119	15,603	24,979
Coal to Dickinson's mills	40,480	43,262	37,025	33,794	45,257
Foxton Lift (locks) traffic	33,425*	40,767	15,482	11,037	9,994
Tring Summit (Lock 45) traffic	295,388*	343,110	281,109	255,125	313,719
Norwood (Lock 90) traffic	355,458†	380,374	361,420	329,708	391,006

Note: The materials to Dickinson's mills at Apsley, Croxley, Home Park and Nash included rags, waste paper and shavings from Paddington with woodpulp, chemicals, china clay and esparto grass from Brentford.
* figure for 1905 † figure for 1906

Whilst relying mainly on outside carriers, John Dickinson & Co operated a few craft of their own. Here their *Lord Nelson* lies below Nash mill bottom lock. *John Dickinson & Co Ltd*

Appendix 3
Lock and Distance Table

This table is based on a detailed survey carried out by Thomas Milner, the company's Northern District engineer, between 18 September 1893 and 18 October 1893, but with subsequent additions. All the locks on the main line of the canal, all the significant branches and docks, all the railway bridges, and some other features are included. Road bridges have generally been omitted unless they are adjacent to a lock, or are relevant to the text.

		Miles	Yards
Braunston: Junction with Oxford Canal & Braunston Branch	[O/D 322.31]	0	0
Braunston Bottom Lock No 1 (rise 6.22ft) & Bridge No 2		0	648
Braunston Gauging Dock		0	682
Braunston Lock 2 (rise 6.60ft) & Bridge 3		0	858
Braunston Lock 3 (rise 5.60ft) & Bridge 4		0	1131
Braunston Lock 4 (rise 6.10ft)		0	1348
Braunston Lock 5 (rise 6.13ft)		0	1477
Braunston Top Lock 6 (rise 4.69ft) & Bridge 5	[O/D 357.65]	0	1631
Braunston Tunnel—west end (Length 2,048yd 2ft)		1	408
Braunston Tunnel—east end		2	697
Welton Feeder enters canal		2	1654
Feeder from Daventry Reservoir enters canal		2	1698
Norton: Junction with the Leicester Line		4	275
Long Buckby Top Lock 7 (fall 8.71ft) & Bridge 11 (Watling Street)		4	639
Long Buckby Lock 8 (fall 8.46ft)		4	1292
Long Buckby Lock 9 (fall 8.81ft)		5	20
Railway Bridge—London Euston to Birmingham line (first crossing)		5	134
Long Buckby Lock 10 (fall 8.86ft) & Bridge 14		5	478
Long Buckby Lock 11 (fall 8.86ft)		5	719
Lock Buckby Lock 12 (fall 9.55ft) & Bridge 15 (Norton Road)		5	974
Long Buckby Bottom Lock 13 (fall 9.50ft)	[O/D 294.90]	5	1148
Norton Feeder enters canal		6	1677
Bridge 22 (Watling Street—second crossing)		7	1147
Junction with Weedon Military Dock (1,156yd) & Brook Street Aqueduct		9	90
Aqueduct over the River Nene		9	370
Church Street Aqueduct		9	683
Stowe Hill Bridge 26 (Watling Street—third crossing)		9	1552
Site of High House Dock & High House Bridge 29 (Stowe Hill Road)		10	1118
Bugbrooke Valley Aqueduct over Horsestone Brook		13	462
Gayton: Junction with Northampton Branch & Bridge 48 (Milton Road)		16	967
Railway Bridge—former Blisworth to Northampton line		16	1318

A cast-iron milepost of 1895.

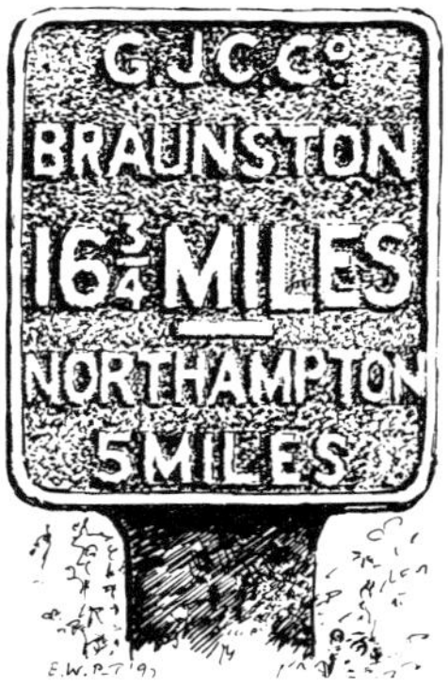

The sign of The Boat at Stoke Bruerne, still a meeting place for canal devotees.

The two-arched bridge at Church lock, No 29.

Railway Bridge—London Euston to Birmingham line (second crossing)		16	1739
Blisworth Tunnel—north end (Length 3075yd 2ft)		18	341
Blisworth Tunnel—south end		19	1657
The Canal Museum (originally Stoke Bruerne Mill & Dock)		20	540
Stoke Bruerne Top Lock 14 (fall 7.46ft) & Bridge 53		20	587
Stoke Bruerne Lock 15 (fall 8.22ft)		20	760
Stoke Bruerne Brick & Tile Works Dock (100yd)		20	892
Stoke Bruerne Lock 16 (fall 7.55ft)		20	1315
Stoke Bruerne Lock 17 (fall 8.17ft)		20	1436
Stoke Bruerne Lock 18 (fall 6.87ft) & Bridge 54 (Stratford Road)		20	1559
Stoke Bruerne Lock 19 (fall 9.20ft)		20	1755
Stoke Bruerne Bottom Lock 20 (fall 8.40ft)	[O/D 239.03]	21	166
River Tove passes under canal		21	591
Solomon's Ornamental Bridge 65 (Road in Cosgrove village)		26	1067
Cosgrove: Junction with the Old Stratford Branch		26	1622
Cosgrove Lock 21 (fall 3.30ft)	[O/D 235.73]	27	670
Wolverton Aqueduct over the River Great Ouse		27	706
Railway Bridges—sidings into Wolverton Works		28	1143
Railway Bridge—London Euston to Birmingham line (third crossing)		29	50
Aqueduct over Grafton Street, Milton Keynes		c29	800
Railway Bridge—former branch line to Newport Pagnell		31	786
Great Linford: former junction with the Newport Pagnell Canal		31	1616
Simpson Aqueduct (over a road)		37	483
Site of Simpson Dock		37	824
Watling Street Bridge (modern A5—fourth crossing)		c37	1100
Fenny Stratford Lock 22 (rise 1.02ft) & Bridge 95 (across lock)		38	717
Railway Bridge—Bletchley to Bedford line		38	798
Fenny Stratford Bridge 96 (formerly Watling Street A5)		38	1087
Fenny Stratford Basin & Dock		38	1140
Water Eaton Aqueduct (over a stream)		39	339
Stoke Hammond Lock 23 (rise 7.42ft) & Bridge 104		41	691
Soulbury Bottom Lock 24 (rise 6.94ft)		42	852
Soulbury Middle Lock 25 (rise 6.74ft)		42	945
Soulbury Top Lock 26 (rise 6.57ft) & Bridge 107	[O/D 264.42]	42	1034
Leighton Lock 27 (rise 6.67ft)		45	703
Site of Leighton Buzzard Dock		46	382
Site of Faulkner's Boatyard		46	545
Site of Railway Bridge—Leighton Buzzard to Dunstable line		46	1174
Ledburn Stream enters canal		47	915
Grove Lock 28 (rise 7.50ft)		47	983
Church Lock 29 (rise 6.83ft) & Bridge 116		48	140
Slapton Lock 30 (rise 7.08ft)		49	1447
Horton Lock 31 (rise 6.75ft) & Bridge 121		50	600
Ivinghoe Bottom Lock 32 (rise 7.08ft)		50	1727
Ivinghoe Top Lock 33 (rise 7.17ft) & Bridge 122		51	238

Seabrook Bottom Lock 34 (rise 7.17ft)		51	1438
Pitstone Brook enters canal		52	198
Seabrook Lock 35 (rise 6.50ft)		52	253
Seabrook Top Lock 36 (rise 6.67ft)		52	596
Railway Bridge—London Euston to Birmingham line (fourth crossing)		52	1064
Mottimore Brook enters canal		53	288
Marsworth Bottom Lock 37 (rise 7.44ft)		53	667
Marsworth Lock 38 (rise 6.90ft)	[O/D 348.18]	53	793
Marsworth: Junction with the Aylesbury Branch		54	295
Startopsend Lock 39 (rise 6.91ft) & Bridge 132 (Ivinghoe Road)		54	587
Marsworth Lock 40 (rise 6.31ft)		54	1099
Railway Feeder North enters canal		54	1190
Marsworth Lock 41 (rise 6.26ft)		54	1290
Marsworth Lock 42 (rise 6.18ft)		54	1446
Marsworth Lock 43 (rise 5.83ft)		54	1590
Marsworth Lock 44 (rise 6.32ft)		54	1738
Marsworth Top Lock 45 (rise 4.98ft)	[O/D 390.97]	55	134
Bulbourne: Junction with the Wendover Branch		55	192
Bulbourne Workshops & Depot		55	629
Railway Feeder South enters canal		56	242
Cowroast Lock 46 (fall 5.97ft) & Bridge 137 (Norcott Hill Road)		58	314
Dudswell Top Lock 47 (fall 6.49ft)		58	1034
Dudswell Bottom Lock 48 (fall 6.90ft) & Bridge 138		58	1351
Northchurch Top Lock 49 (fall 6.77ft) & Bridge 139		59	935
Northchurch Lock 50 (fall 6.82ft)		59	1604
Northchurch Lock 51 (fall 6.53ft)		60	598
Northchurch Bottom Lock 52 (fall 6.88ft)		60	738
Broadwater Lock 53 (fall 5.83ft) & Lower Kings Road Bridge		60	1404
Berkhamsted Boat Yard Dock (now Bridgewater Boats)		61	199
Berkhamsted Lock 54 (fall 5.67ft) & Bridge 142 (Ravens Lane)		61	300
Berkhamsted Lock 55 (fall 5.37ft)		61	465
Branch of River Bulbourne enters canal		61	500
Topside Lock 56, Berkhamsted (fall 8.04ft)		61	1502
Bottomside Lock 57, Bourne End (fall 8.49ft)		62	237
Sewerage Lock 58, Bourne End (fall 6.08ft)		62	710
Bourne End Bottom Lock 59 (fall 6.75ft)		62	1699
Junction with channel to Bourne End Mill		63	91
Winkwell Top Lock 60 (fall 7.13ft)		63	416
Winkwell Bottom Lock 61 (fall 6.43ft)		63	797
Site of Cranstone's Gravel Pit Dock		63	841
Railway Bridge—London Euston to Birmingham line (fifth crossing)		63	1188
Boxmoor Top Lock 62 (fall 6.70ft)		63	1713
Fishery Lock 63, Boxmoor (fall 7.11ft) & Bridge 149 (Fishery Lane)		64	659
River Bulbourne passes under canal in culvert		64	1038
Boxmoor Bottom Lock 64 (fall 7.10ft)	[O/D 263.91]	65	99

Heavy traction engines posed a threat to canal bridges, and this weight restriction sign was erected at a bridge approach.

A hand-operated swing bridge at Winkwell.

The double-arched bridge, No 154, at Apsley to pass floodwater off the Gade.

Site of railway bridge—former Hemel Hempstead to Harpenden line	65	175
River Gade enters canal & junction with channel to Two Waters Mill	65	275
Junction with original line of canal avoiding the mills	65	1100
Apsley Top Lock 65 (fall 4.61ft) & Bridge 153	65	1124
Apsley Lock 66 (fall 5.21ft)	65	1410
Junction with channel to Frogmore Mill	65	1460
Apsley Bottom Lock 67 (fall 6.28ft)	65	1711
Nash Mill Top Lock (fall 5.90ft) & channel to Nash Mill	66	784
Nash Mill Bottom Lock 69 (fall 6.27ft) & Bridge 155	66	1032
Railway Bridge—London Euston to Birmingham line (sixth crossing)	66	1530
Approximate site of junction with original canal	66	1550
Junction with channel to Kings Langley Mill	67	968
Kings Langley Lock 69A (fall 8.77ft)	67	1160
Junction with channel to Home Park Mill	68	198
Home Park Mill Lock 70 (fall 5.21ft)	68	298
M25 Motorway Viaduct	c68	1020
Home Park Farm lock 71 (fall 6.91ft)	68	1277
Junction with channel to Hunton Bridge Mill	69	545
Hunton Bridge Top Lock 72 (fall 5.61ft)	69	711
Hunton Bridge Bottom Lock 73 (fall 5.59ft)	69	950
Lady Capel's Lock 74, Watford (fall 5.34ft)	70	211
Grove Ornamental Bridge 164 (private drive)	70	753
Junction with channel to Grove Mill	70	968
Cassiobury Park Top Lock 75 (fall 4.63ft)	70	1727
Cassiobury Park Lock 76 (fall 5.39ft)	71	152
Ironbridge Lock 77, Watford (fall 7.36ft) & Bridge 167	71	952
Cassiobury Park Bottom Lock 78 (fall 8.97ft)	72	415
Railway Bridge (high level)—Metropolitan line to West Watford	72	570
Railway Bridge—branch line to Croxley Green	72	860
Common Moor Lock 79, Croxley Green (fall 9.39ft)	73	240
Railway Bridge—London Marylebone to Aylesbury line	73	1318
Lot Mead Lock 80 (fall 6.24ft) [O/D 156.23]	73	1530
Railway Bridge—former branch line to Rickmansworth	73	1664
River Colne enters canal	74	29
Batchworth Lock 81, Rickmansworth (fall 6.69ft)	74	1123
Batchworth: Junction with the Rickmansworth Branch	74	1162
Former junction with Bury Grounds Branch	74	1281
Junction with channel to Batchworth Mill	74	1420
Site of Frogmoor boatyard and dock	74	1576
Stockers Lock 82 (fall 5.14ft) & Bridge 175	75	670
City Dues Stone (Coal Duty Marker)	75	1392
Springwell Lock 83 (fall 7.92ft) & Springwell Bridge 176	76	142
Copper Mill Lock 84, Harefield (fall 5.84ft)	77	258
Junction with the Troy Branch	77	828
Black Jack's Lock 85, Harefield (fall 3.70ft) & Bridge 178	77	1349

Widewater Lock 86, Harefield (fall 8.04ft) & Bridge 180	78	1561
Harefield Moor Dock (now Harefield Cruising Club)	78	1672
Railway Bridge—London Marylebone to High Wycombe line	79	1140
Aqueduct over Frays River	80	772
Denham Deep Lock 87 (fall 11.06ft)	80	782
Junction with channel to King's Mill & Dock (220yd)	81	477
Uxbridge Lock 88 (fall 4.57ft) & Bridge 184	81	622
River Frays Dock (220yd—now Denham Marina)	81	748
Osborne Stevens Dock	81	814
Site of Way's Wharf & Dock (330yd)	81	1080

A motor boat about to enter the north end of Blisworth tunnel in the summer of 1955.
Alan Faulkner

Uxbridge Boatyard Dock & Dolphin Bridge 186 (Slough Road)		82	132
Cowley Lock 89 (fall 6.42ft) & Frays River Aqueduct	[O/D 96.85]	83	74
Site of Cowley Peachey Dock (350yd)		83	506
Cowley Peachey Packet Boat Dock		83	1253
Cowley Peachey: Junction with the Slough Branch		83	1595
Aqueduct over the River Pinn		83	1675
Site of railway bridge—former branch line to Uxbridge		83	1699
Site of Yiewsley Dock (Onslow Mills)		84	292
Site of Holland's Dock (Bentinck Dock)		84	647
Site of Otter Dock (Waterworks Dock) (1,845yd)		84	1017
Site of Cooper's Dock (Dutton's/Houghton's Boatyard Dock) (175yd)		84	1594
Site of Eastwood's Dock (Liddall's/Sabey's Dock) (610yd)		85	71
Site of Rutter's Dock		85	79
Site of Stockley Dock (350yd)		85	341
Site of Dawley Dock (Wallington's Dock) (320yd)		85	638
Pocock's Dock (Broad's/Starveall Dock) (1,120yd)		85	1194
Site of Maynard's & Odell's Dock (300yd)		85	1741
Site of Pinkwell's Brickyard Dock		86	49
Site of Odell's Short Dock (McAlpine's Dock)		86	291
Site of Wilshin's Dock (Shackle's Brickyard Dock) (200yd)		86	698
Site of Printing House Dock (100yd)		86	1350
Site of Great Western Railway Dock (165yd)		87	32
Chair Dock, Hayes (Shackle's Dock) (100yd)		87	82
Railway Bridge—London Paddington to Bristol line		87	264
Aqueduct over Yeading Brook		87	799
Bulls Bridge: Junction with the Paddington Branch		87	993
Bulls Bridge Boatyard Dock		87	1020
Site of Iszard's Dock (Mead's Dock)		87	1347
Site of junction with North Hyde Military Dock (1,495yd)		88	788
Site of Victoria Dock (Passenham Dock) (580yd)		88	1163
Adelaide Dock (Gibson's Dock) (160yd)		88	1453
Site of Passmore's Dock (Old Ruislip feeder) (1,080yd)		89	673
Maypole Dock (Otto Monsted Dock) (780yd)		89	856
Norwood Top Lock 90 (fall 7.80ft) & Glade Lane Bridge 204		89	1136
Norwood Bottom Lock 91 (fall 7.91ft)		89	1411
Windmill Bridge 205 (road over canal over railway)		90	28
Hanwell Top Lock 92 (fall 7.86ft)		90	253
Hanwell Lock 93 (fall 10.00ft)		90	443
Hanwell Lock 94 (fall 10.16ft) & site of Hanwell Asylum Dock		90	574
Hanwell Lock 95 (fall 9.33ft)		90	702
Hanwell Lock 96 (fall 7.83ft)		90	818
Hanwell Bottom Lock 97 (fall 7.25ft)	[O/D 28.71]	90	941
River Brent enters canal		90	1015
Site of Doctor Hume's Dock		90	1443
Osterley Lock 98 (fall 5.53ft)		91	642

The entrance to Adelaide Dock, spanned by the towpath bridge.

Railway Bridge (high level)—Piccadilly tube line to Heathrow		91	1209
Clitheroe's Lock 99 (fall 7.58ft)		92	327
Railway Bridge—London Waterloo to Hounslow line		92	1100
River Brent backwater to part of Brentford Depot		92	1232
Fellows, Morton & Clayton Dock, Brentford		92	1483
Brentford Gauging Locks 100 (fall 5.30ft)	[O/D 10.30]	92	1513
River Brent backwater—Brentford Bridge Boatyard		92	1558
Railway Bridge—former branch line to GWR's Brentford Dock		92	1668
Site of Brentford Weigh Dock		93	229
Railway Bridge—former branch line to GWR's Brentford Dock		93	309
Workhouse Dock		93	436
Doctor Johnson's Dock		93	556
Thames Locks 101 (tidal—fall varies from 10.30ft to nil)		93	617
Junction of the canal with the River Thames		93	986

Gate mechanism at Brentford gauging lock, No 100.

GRAND JUNCTION BRANCHES

LEICESTER LINE

Grand Union Section: Norton Junction to Foxton. 23 miles 231yd. Seven locks at Watford rise 54.06ft (Watford Summit Pound O/D 411.71). Ten locks at Foxton (in two flights of five) fall 75ft. Section includes Crick Tunnel 1,527½yd long, and Husbands Bosworth Tunnel 1,170½yd long. 7ft beam.

Old Union Section: Foxton to Leicester. 18 miles 168yd. Twenty-four locks between Kibworth and Leicester fall 160.83ft. Section includes Saddington Tunnel 881½yd long. 14ft beam.

Aylesbury: 6 miles 317yd. Sixteen locks fall 94.64ft to Aylesbury. 7ft beam.
Braunston: 230yd to Braunston wharf. Level. 7ft beam.
Buckingham: 8 miles 1720yd. Two locks rise 13.75ft to Buckingham. 7ft beam.
Foxton Lift: Top Arm 350yd. Bottom Arm 180yd. Both level. 14ft beam.
Market Harborough: 5 miles 895yd. Level. 14ft beam. (Originally part of the Old Union main line)
Northampton: 4 miles 1311yd. Seventeen locks fall 107.10ft to Northampton. 7ft beam.
Old Stratford: 1 mile 533yd. Level. 14ft beam.
Paddington: 13 miles 979yd. Level. 14ft beam.
Rickmansworth: 536yd. One lock rises 2.66ft. 14ft beam.
Slough: 4 miles 1626yd. Level. 14ft beam.
Troy: 1050yd. Level. 14ft beam.
Welford: 1 mile 1110yd. One lock rises 5ft. 7ft beam.
Wendover: 6 miles 1352yd. Level, but with stop lock at Tringford. 14ft beam.

Watford locks near the junction of the Leicester line with the Grand Junction, rebuilt in 1902.

Grand Junction Terminology

Barge: A boat with a beam of 14ft that mainly frequented the southern reaches of the canal up to Berkhamsted. Such vessels were usually unpowered, and were towed by horses, or latterly by small tractors.

Butty (boat): A narrow boat working in company with another.

Compensation water: Water supplied from reservoirs in dry seasons to millers and others to replace water abstracted from rivers and streams at other times for the use of the canal; eg, Aldenham reservoir was built to supply compensation water to the millers on the river Colne.

Culvert: A conduit or underground drain to carry water under the canal. Several streams had to be culverted underneath the bed of the Grand Junction when the canal was being built.

Dock: A short branch, usually serving a particular factory or wharf; eg, the Maypole Dock at Southall, which served the Maypole margarine works.

Construction work on Maypole Dock at Southall in 1913.
National Waterways Museum

Feeder: A channel through which water was supplied to the canal by gravity or natural flow. In most cases these would be concentrated at the summit levels; eg, the Tring feeder which supplies water to the Wendover branch, and hence to the Tring summit.

The Three Bridges at Southall where road, canal and railway intersect.

Fly boat: A boat working day and night, originally using relays of horses.

FMC: Fellows, Morton & Clayton Limited. The company was based on a business founded in 1837, was incorporated in 1889, and eventually went into voluntary liquidation in 1948.

Gauging: A process to ascertain the weight of cargo by measuring the craft's freeboard, and relating the measurement to a special table prepared in advance by the canal company for each individual craft.

GJWWC: Grand Junction Water Works Company. The company was incorporated in 1811 and remained in being until 1902.

GUC: Grand Union Canal. This was the original canal opened in 1814 between Norton and Foxton. From 1929 the name was adopted by the Regent's Canal & Dock Company, following its acquisition of the Grand Junction and the Warwick Canals, to signify its wider purpose.

GWR: Great Western Railway.

Heading: An underground tunnel, or culvert, built to allow water to be passed between different places; eg, the Wilstone heading links the three Wilstone reservoirs with the Tringford pumping station. A cross heading was a branch off the main heading, such as used in the construction of Blisworth tunnel.

Inclined plane: A device on wheels which lifts boats from one level of a canal to another, without using locks. The last plane to be constructed in this country, and one of the best known, was the Grand Junction's installation at Foxton in Leicestershire.

Invert: An inverted arch of brickwork or masonry used to form the bottom of locks, bridges and tunnels where additional strength is required to withstand lateral or upward pressures in the ground. Blisworth tunnel has an invert throughout its entire length.

K&AC: Kennet & Avon Canal.

L&BR: London & Birmingham Railway, which in 1846 was merged to become part of the London & North Western Railway.

L&NUC: Leicestershire & Northamptonshire Union Canal, often more commonly known as the Old Union Canal.

The lock-keepers cottage at Dudswell top lock, No 47.

L&NWR: London & North Western Railway.

Legging: A method to propel unpowered craft through tunnels which had no towing path. The leggers lay on planks, fixed across the bows of the boat, and pushed with their feet against the tunnel walls. Legging was standard practice at Blisworth and Braunston tunnels for many years.

Lock: An enclosed chamber, with gates at each end, by means of which a boat can be raised, or lowered, from one level of canal to another.

Locks: On the Grand Junction the capacity of reservoirs and feeders was usually computed in "locks". A lock measure was the equivalent of 56,000 gallons of water.

Long Level: The level stretch of canal running from Cowley lock to Norwood lock and including all the Paddington branch, the Slough branch, and many shorter docks. The level acted as an extended reservoir for the busy locks down to Brentford, and on the connecting Regent's Canal.

Narrow boat: A boat measuring some 70ft long by 7ft wide that was used extensively on the canal. This was the standard canal vessel used by the long-distance carriers.

New Pit: The second pumping well shaft at Tringford, sunk in 1835/36.

Northern Engines: A series of pumping stations between Fenny Stratford and Marsworth, designed to lift water back to the Tring reservoirs.

Old Pit: The original pumping well shaft at Tringford, sunk in 1816/17.

Portal: The entrance to a tunnel; eg, the northern portal of Blisworth tunnel, which was rebuilt in 1902/03.

Pound: The stretch of canal between two locks; eg, the Blisworth pound, almost fifteen miles in length, from the tail of the Long Buckby locks to the head of the Stoke Bruerne locks.

Puddle: Clay, worked up with water into a smooth consistency, and spread on to the bottom and sides of the canal to make it watertight.

Side pond: A water-saving device at locks whereby part of the water was discharged into an intermediate chamber alongside the main lock, and then re-used to part fill the lock for the next operation.

Stoppage: A temporary closure of the canal for repairs. Stoppages provided an ideal time to see boats, which were held up until the work was complete.

Summit (level): The highest level of the canal and the one that had to be kept fully supplied with water. The Grand Junction's summits were at Braunston and Tring, with another on the Leicester line.

Toll: The charge levied on a carrier for the use of the canal. The income from tolls formed one of the main sources of income for the Grand Junction.

Top lock: The uppermost lock in a flight; eg, Braunston top lock No 6 admits craft on to the Braunston summit level.

Towpath: The path beside the canal used by the towing horses. There is a towpath virtually throughout the Grand Junction, except through the tunnels.

Turnover bridge: A bridge taking the towpath from one side of the canal to the other. These occur regularly through the Grand Junction system.

W&BC: Wilts & Berks Canal.

Based on the top lock of the Marsworth flight, this pictorial diagram shows the main features of a Grand Junction lock

Work in progress at Bulbourne in April 1903 to waterproof a crucial section of the Tring summit level. This involved building brick walls on both sides of the canal and lining the bed with concrete. *W.H. Walker & Brothers Ltd.*

Weir: A barrier, or dam, constructed to regulate water levels. All the Grand Junction reservoirs have such a controlling weir, over which the water discharges automatically once the level is reached. Many pounds have a similar overflow weir discharging into a convenient local stream.

Wide boat: A boat with a length of 70ft and a beam from 9ft to 12ft, dimensions between those of a barge and a narrow boat. Such craft were seen in considerable numbers mainly on the lower reaches of the Grand Junction although they could, and did, trade as far north as Braunston.

WJC: Western Junction Canal (project).

WUC: Western Union Canal (project).

Index

Illustrations in bold type

Blisworth tunnel tug *Spider*,
1870-1928

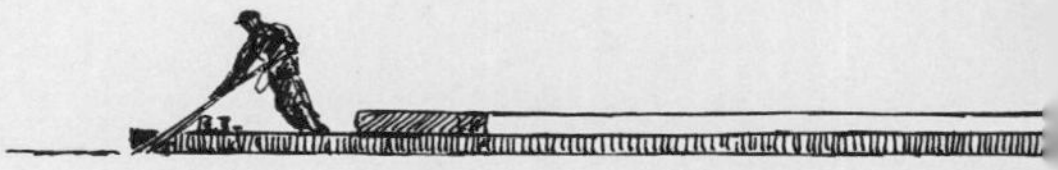

Directors' launch *Kingfisher*, 1928

Icebreaker *Paddington*, built by
Walkers at Rickmansworth in 1919